ALSO BY ALEX BELLOS

Puzzle Me Twice

Perilous Problems for Puzzle Lovers

Can You Solve My Problems?

Puzzle Ninja

Here's Looking at Euclid

The Grapes of Math

Futebol: The Brazilian Way of Life

WITH EDMUND HARRISS

Patterns of the Universe:
A Coloring Adventure in Math and Beauty

Visions of the Universe:
A Coloring Journey through Math's
Great Mysteries

THE LANGUAGE LOVER'S PUZZLE BOOK

Perple_ing Le_ical
Patterns to Unmi_
and Ve_ing Synta_
to Outfo_

ALEX BELLOS

THE EXPERIMENT
NEW YORK

THE LANGUAGE LOVER'S PUZZLE BOOK: *Perplexing Lexical Patterns to Unmix and Vexing Syntax to Outfox*
Copyright © 2020 by Alex Bellos

Originally published in the UK by Guardian Faber, an imprint of Faber & Faber Ltd, in 2020. First published in North America in revised form by The Experiment, LLC, in 2021.

The Experiment, LLC
220 East 23rd Street, Suite 600
New York, NY 10010-4658
theexperimentpublishing.com

THE EXPERIMENT and its colophon are registered trademarks of The Experiment, LLC. Many of the designations used by manufacturers and sellers to distinguish their products are claimed as trademarks. Where those designations appear in this book and The Experiment was aware of a trademark claim, the designations have been capitalized.

The Experiment's books are available at special discounts when purchased in bulk for premiums and sales promotions as well as for fund-raising or educational use. For details, contact us at info@theexperimentpublishing.com.

Library of Congress Cataloging-in-Publication Data

Names: Bellos, Alex, 1969- author.
Title: The language lover's puzzle book : perple_ing le_ical patterns to
    unmi_ and ve_ing synta_ to outfo_ / Alex Bellos.
Description: New York : The Experiment [2021] | Includes index. | Summary:
    "100 wonder-filled word puzzles that thrill and tantalize with the
    beauty, magic, and weirdness of world language"-- Provided by publisher.
Identifiers: LCCN 2021033410 (print) | LCCN 2021033411 (ebook) | ISBN
    9781615198047 (paperback) | ISBN 9781615198054 (ebook)
Subjects: LCSH: Word games. | Language and languages.
Classification: LCC GV1507.W8 B44 2021 (print) | LCC GV1507.W8 (ebook) |
    DDC 793.734--dc23
LC record available at https://lccn.loc.gov/2021033410
LC ebook record available at https://lccn.loc.gov/2021033411

ISBN 978-1-61519-804-7
Ebook ISBN 978-1-61519-805-4

Cover design by Jack Dunnington
Text design by carrdesignstudio.com
Author photograph by Michael Duerinckx

Manufactured in the United States of America

First printing November 2021
10 9 8 7 6 5 4

To my polyglot parents

Contents

Introduction

My name is Alex and I'm a language lover.

I'm also a puzzle person. You could call me a conundrum connoisseur. Indeed, for the last five years I've been writing a puzzle column in *The Guardian*. What I love about puzzles is the way they provide a double hit: They present a fun challenge, while opening your eyes to something new. A good puzzle gives you the thrill of achievement and the joy of discovery at the same time.

The brainteasers in this book celebrate language, and languages. You will find enigmas about English, a riddle about Rarotongan, and a stumper about stenography, as well as dozens of other problems about exotic tongues and mysterious alphabets. You'll decipher Egyptian hieroglyphs, translate puns from Dutch, and deduce the days of the week in Swahili. You will learn how to count in Manx, how to write in Chinese, and how to speak in silence to an eleventh-century monk. Each puzzle reveals something curious about language, about culture, or about science. Most of it will be new to you.

Here's an example, to give you an idea of what to expect.

The following eight phrases are in Japanese (transcribed into the Latin alphabet), followed by their translations:

boru niko	two balls	*ashi gohon*	five legs
tsuna nihon	two ropes	*ringo goko*	five apples
uma nito	two horses	*sara gomai*	five plates
kami nimai	two sheets of paper	*kaba goto*	five hippos

What is the Japanese for nine cucumbers?

a) *kyuri kyuhon* c) *kyuri kyuhiki*

b) *kyuri kyuko*   d) *kyuri kyuto*

Think about it for a while. We'll get to the solution soon enough. But you can already see that Japanese speakers do not count objects the way English does.

In 1965, academics at Moscow State University organized a competition that aimed to get secondary school students excited about languages. It was modeled on the "mathematical olympiad," a math competition that was attracting many young students to the subject. The "linguistics olympiad" consisted of brainteasers about language and languages that were intended to be fun, enticing, and curious. These contests have since inspired linguists around the world to start their own competitions using similar types of questions. The UK Linguistics Olympiad was inaugurated in 2010.

Two years ago, a reader of my *Guardian* column tipped me off about the linguistics olympiads. I was entranced! The problems blew my mind. They were beautiful and varied, twisting my brain in ways it had never been twisted before. At the same time, they transported me across the world and introduced me to languages and writing systems that I had never even heard of. When I published some of these puzzles in my column, my readers loved them too.

This book is a compilation of my favorite olympiad problems. They assume zero knowledge of linguistics, and no competence in a language other than English. They can be solved with a mix of mathematical reasoning, linguistic intuition, outside-the-box thinking, and bits of general knowledge. If you've ever solved a sudoku, completed a crossword, or scrutinized a Scrabble board, then these puzzles will get your juices flowing.

Around 7,000 languages are currently spoken in the world. You will encounter more than 50 of them in the following pages. I've also included problems about more than a dozen ancient tongues, and several invented languages.

Many of the puzzles are code-breaking challenges in which your task is to deduce what is interesting, or unusual, about the language in question. The easier ones concern languages similar to English. The further from English we get, the harder the problems become. I've chosen indigenous languages from every continent (except Antarctica, which doesn't have any) that display all sorts of phenomena, to showcase, as much as I can, the amazing linguistic diversity in the world. The puzzles require you to process the language in your head, so they also give you the thrill of thinking the way that speakers of that language do. For a moment you leave your couch for ancient Babylon, or Papua New Guinea, or Timbuktu.

Several puzzles introduce unfamiliar alphabets and writing systems. There is something magical about looking at a script you cannot read, and then trying to decipher it. At first, all you see is the elegance of the shapes, artistic expressions of a culture that is not your own. Gradually, patterns reveal themselves. The

puzzles I have included give you just enough information to decipher them. The joyous moment when it all fits together is an
echo, perhaps, of the moment you learned to read as a child.

I've arranged the puzzles by theme, and provided context
and background that tell wider stories, such as the development of writing, the evolution of English, the birth of numerical notation, and the history of invented languages. We
will meet incredible people, such as Sequoyah, who invented
the first Cherokee writing system, and Louis Braille, who devised his tactile alphabet for blind people when he was only
15 years old. I will explore how the study of language—
and its connections with geography, history, sociology, and
psychology—contributes to a deeper understanding of the
human experience.

Our travels take us back in time and around the world.
They also take us to the future. Linguistics has always been
a poor relation among the sciences, lacking the romance of,
say, astrophysics, the fashionableness of genetics, or the nerd
cool of pure math. Yet the field is perhaps at the most exciting and relevant point it has ever been, thanks to the growth
of language-based tech such as voice recognition, machine
translation, and natural language processing. The next great
leaps in artificial intelligence (AI) will rely on breakthroughs
in our understanding of language.

The book begins with a chapter on language and technology.
We'll have fun with Google searches, internet slang, and some
extremely efrimious bots. (You'll understand what I mean when
we get there.) The problems in the first chapter are almost all in
English, which means they provide something of a warm-up for
the rest of the book. We'll get into the swing of discovering how
languages work by looking at English before venturing further
afield.

The puzzles are not presented in order of difficulty. Some are particularly fiendish, because they require your brain to work in unfamiliar ways. Once you know the answer, of course, the hardest problems turn out to be almost trivially easy. Languages are designed for communication, not obfuscation, even if to nonspeakers they can appear convoluted or complex. I include some that were deliberately invented to be as simple as possible, and these result in some of the toughest puzzles in the book! (Before each chapter I also include some quick-fire questions to get you in the mood.)

Now back to those Japanese cucumbers. Let's solve the problem slice by slice.

Cast your eyes across the Japanese and English words. The most noticeable pattern is that the English phrases all have the word "two" in them, but the Japanese phrases share no repeated whole word. We can deduce that "two" has no single translation in Japanese. Likewise, from the second list we can deduce that "five" has no single translation either.

Yet the Japanese phrases do exhibit some kind of pattern. Looking closely, we can see that the element *ni-* is repeated in every phrase in the first list, and the element *go-* is repeated in every phrase in the second. Since we know that every phrase in the first list contains the idea of "two," and every phrase in the second contains the idea of "five," it seems likely that *ni-* means "two" and *go-* means "five."

This inference raises the question of what the other parts of the *ni-* and *go-* words might mean. You may have spotted that the *ni-* and *go-* words are followed by the same four suffixes: *-ko*, *-hon*, *-to*, and *-mai*.

Let's arrange the items in the phrases according to these suffixes.

-ko	balls, apples
-hon	ropes, legs
-to	horses, hippos
-mai	sheets of paper, plates

Now we're getting to the business end of the solution. To get to this point we've used visual pattern recognition and deductive logic. Now our brains must switch to considering the meanings of the words. In what way are balls and apples similar, or ropes and legs, or horses and hippos, or paper and plates?

The answer is that when you count things in Japanese, you cannot use numbers by themselves. You must also describe the type of thing you're counting by adding a suffix to the number word. To wit:

-ko	round things
-hon	long, thin things
-to	big animals
-mai	flat, thin things

A cucumber is a long, thin thing, so takes the suffix *-hon*. The Japanese for "nine cucumbers" is therefore *kyuri kyuhon*.

The pleasure here is not just the satisfaction of solving a puzzle, but also discovering something fascinating about another language. The Japanese system of number suffixes is astonishing when you learn about it for the first time. Counting is so simple, yet they make it so difficult! Japanese has hundreds of number suffixes. There is one for floors (*-kai*), another one for books (*-satsu*), another one for cars (*-dai*), another one for swordstrokes (*-furi*), and so on.

Hold your breath, however, before you ridicule the Japanese language for needless overcomplication. English often does exactly the same thing. We do not say "two cattles," "three breads," or "four papers"; we say "head of cattle," "loaves of bread," and "pieces" or "sheets of paper." Our language is really just as perplexing as Japanese is. So linguistics puzzles actually deliver three times over. You are forced to use your wits, you discover something surprising about another language, and, finally, you learn something about your own language too.

Most important, rather than simply being told of a particular linguistic phenomenon, you get the pleasure of working it out for yourself. You flex your linguistic and logical brain, and learn something new about the world.

LINGO (B)(I)(N)(G)(O)

Vocab Test A–L

Guess the correct meaning of the word.

1. ABSCONCE
   a) Do a runner
   b) Conceal
   c) Drink a yard of ale as a result of losing a bet
   d) Lantern used in monasteries

2. BANDOLINE
   a) Extinct instrument resembling a ukulele
   b) Mucilaginous preparation made from quince seeds, used for setting hair
   c) Hair band
   d) Early form of sticky tape

3. BARDO
   a) French film star
   b) Lock-in
   c) Troubador
   d) State between death and rebirth (in Buddhism)

4. BICHON FRISÉ
   a) Small breed of dog
   b) French insult
   c) Grated carrots in nouvelle cuisine
   d) 1920s hairstyle

5. BULSE
   a) Dust on a threshing floor
   b) Thin porridge
   c) Bag for carrying diamonds in
   d) Reindeer droppings

6. CHIMANGO
   a) Brazilian monkey
   b) Male character in South American show akin
      to Punch and Judy
   c) Lychee-like fruit in northern South America
   d) South American falcon

7. CRUBEEN
   a) Rough girl
   b) Rough drink
   c) Pig's feet as food
   d) Hovel

8. DEMIURGE
   a) Half-formed wish
   b) Encourage, but weakly
   c) Tyrant's henchman
   d) Deity responsible for creating and running
      the universe

9. EPERGNE
   a) Royal pardon
   b) Fencing move
   c) Ornamental centerpiece for dining table
   d) Hungarian strawberry cake beloved of Empress Sisi

10. FLAGITIOUS
    a) Beanlike
    b) Foppish
    c) A little too fond of the whip
    d) Villainous

11. GUAJIRO
    a) Muleteer
    b) Custard apple
    c) Cuban agricultural worker
    d) Man about town

12. GWYNIAD
    a) Welsh epic poem
    b) Endangered whitefish native to Llyn Tegid
    c) Snow White
    d) Snowdonia

13. HIRAGANA
    a) Japanese artificial flower ornament
    b) Japanese national holiday
    c) Japanese good-luck token
    d) Japanese system of syllabic writing

14. KUDZU
    a) Perennial climbing plant
    b) Japanese ceremonial headgear
    c) Fermented soya dish
    d) Western Chinese antelope

1
Ok-Voon
Ororok Sprok

COMPUTER TALK

The information age is a time of plenty for language lovers (IMHO). Emojis. Alexa. Words with Friends. Later in this chapter we'll be exploring the challenges computers face in understanding language, but we'll start by celebrating some word games the internet allows us to play. For example, have you ever tried Googling unusual word pairs to discover the contexts in which they appear? I have, obviously. For the opening puzzles, however, you are forbidden from going anywhere near a computer.

1

ODD COUPLES

For each of the pairs of words shown below, provide a
grammatical and meaningful sentence in which they appear.
In each case the words must appear consecutively in the
order in which they are shown, with no punctuation marks
between them.

a) could to   d) the John
b) he have   e) that than
c) that that

One level up from Google's basic word search is the Ngram Viewer. Search for a word with this tool and it will give you a timeline of that word's frequency of use between the years 1800 and 2008, based on how often it appears in the millions of books in Google's digital archive.

As you can imagine, the Ngram Viewer is a rabbit hole that can take hours out of your life, especially when you discover its advanced features, such as plugging in a word to reveal which other word follows it most frequently. For example, the graph below shows how the nouns most likely to be paired with "puzzle" have changed over the course of the past two centuries, based on Google's database of British English texts from that period. ("Antony" was most popular in 1800, and the others are the three most popular now.)

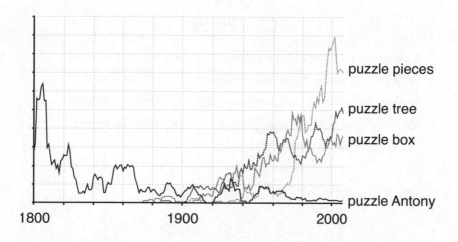

Puzzled as to the context of "puzzle Antony"? Shakespeare scholars will be smug. "Your presence needs must puzzle Antony" is a quote from *Antony and Cleopatra*.

Can you deduce why the second-most-common noun to follow "puzzle" in 2000 was "tree"? (The answer is in the solution to the next question.)

In a word chain puzzle, the setter provides two words, and the solver must find a word that fits between them such that each consecutive pair is a familiar phrase. For example, the solution to "CANDY _ _ _ CHART" is "BAR," since "CANDY BAR" and "BAR CHART" are familiar pairs.

The following problem contains only Ngram-verified word chains.

2

ICE CHEESE

Complete the word chains below. For any word, the word following it is the noun that followed it most frequently in 2008, according to Google's archive of books in British English. (The missing words are all in their singular forms.) Each blank represents a letter.

CREDIT _ _ _ _ GAME

ICE _ _ _ _ _ CHEESE

BEACH _ _ _ _ _ PRICES

COUCH _ _ _ _ _ _ CHIP

BOWLING _ _ _ _ _ _ _ _ FOOD

SALMON _ _ _ _ _ _ _ _ _ _ TRIP

In the next set of chains, each word is followed by the noun that follows it most often in Google's collection of all English books (including American English). I have given you a HINT (and thrown you a BONE) to help you on your way.

SPACE _H_ _ _ _ _ _   B_ _   _ _ _ _ _   SIGN

CHEESE _ _ _ _ _ I _ _   _ _ O _   _ _ _ _ _ _   PLAN

EYE _ _ N _ _ _ _   _ N _ _ _ _ _ _ _ _ _   TECHNOLOGY

COWBOY _ _ _ T   _ _ _ _ _   _ _ _ E   DEPARTMENT

One of the main goals of artificial intelligence (AI) is to create a computer that can communicate perfectly in a human language. This challenge is one of the hardest in AI, as anyone who has shouted in frustration at Siri or Alexa knows full well. We can make robots that are more dextrous than humans, that can see better than humans, and that can compute faster than humans. But no one can yet train a computer so it has anywhere near the level of language understanding a child has.

Here's a glimpse of why. Consider these two simple sentences, which differ by a single word:

"The book did not fit in the box; it was too big."

"The book did not fit in the box; it was too small."

To humans, it is obvious that the "it" in the first sentence refers to the book, and the "it" in the second sentence refers to the box. Computers, however, can't make the switch. Not at the moment, anyway. The tech industry has whole departments of linguistics PhDs trying to solve this type of "natural language" problem. The next big leaps in AI depend on whether or not they succeed.

In the above example, the same word, in the same position, refers to two different things. Likewise, many words have subtle differences in meaning that depend on context, a feature of language that creates nightmarish challenges for computer programmers but which makes for some delightful puzzles.

3

IT STARTED TO RAIN

In each of the three paragraphs below, a single word is missing. In each case, the same word, with subtle variations in meaning, can be used to fill all the blanks in its respective paragraph. What is the missing word in each paragraph?

a) Sarah's venture capital firm bought a _____ and began to fire some of its staff. On her way home she picked up a _____ to read, but it started to rain, so she used the _____ to keep herself from getting wet.

b) I _____ that this is an exciting time in my academic journey, but sometimes I _____ scared of the unforeseen future. Anyway, I need to _____ approval for my vacation. Then, I have to _____ to the airport to pick up my sister. Do you want me to _____ some drinks for you before I leave?

c) This is a _____ time for Matt to upgrade the walls of his house. But he needs a _____ builder for this purpose. His _____ friend Adam, who is also a _____ person, should be able to help him in this regard. They will have a _____ time this summer.

Award yourself a bonus point if you can come up with an alternative word or phrase to put in each blank space that does not change the meaning of the sentence.

To err is human. When it comes to language understanding, however, to err is particularly nonhuman. The next problem simulates what happens when a computer translating a piece of text gets the wrong end of the stick.

4

THE BAD TRANSLATION

In the text below, 20 words have been substituted for alternatives that share a meaning with the original word, but which are incorrect in this context.

For example, in the opening line, the word "cross" has been replaced by "angry," because "cross" and "angry" share the same meaning. In this context, however, "angry-legged" is nonsense.

Find the other 19 words that are incorrect translations, and guess the original words they replaced.

Annie Jones sat angry-legged on her Uncle John's façade porch, her favorite rag doll clutched under one supply. The deceased afternoon sun polished through the departs of the giant oak tree, casting its flickering ignite on the cabin. This entranced the child, and she sat with her confront changed upward, as if hypnotized. A stabilize hum of conversation flowed from inside the cabin.

"Ellen, I'm really happy that you arrived to church with us today. Why don't you spend the night here? It's buying awfully deceased and it will be dark ahead you construct it house."

"I'll be thin, Sally," replied Annie's mother. "Anyhow, you know how Steve is about his supper. I departed plenty for him and the boys on the support of the stove, but he'll want Annie and me house."

Imagine an app designed to read pieces of electronic text. The app might, say, trawl through thousands of text messages to look for, say, clues in a murder inquiry. The app's first job is to identify each word correctly. In a world where everyone was a perfect speller, this would be straightforward. But people make *mistaykes*. And when it comes to text messages and social media slang, people misspell deliberately, to express emotions such as, for example, surprise and amazement. Say whaaaaat?!

How might you teach a computer that "whaaaaat" means "what" without listing every possible deliberate misspelling of "what"? One method, commonly used by computer scientists, is to create a formula, called a "regular expression" or regex, that provides a template for how a word might transform. A regex uses letters combined with the following symbols:

? The previous unit can occur zero or one time

* The previous unit can occur zero or more times

+ The previous unit can occur one or more times

(A unit can be a letter, or it can be a string of letters contained in brackets. If a unit is not followed by a symbol, the unit must appear exactly once.)

The regex WH?A+T* generates *what*, *wat* (because H can occur zero times), *wha* (because T can occur zero times), *whaaaat* (because A can occur more than one time), and other words including *wa*, *whaa* and *whaat*. On the other hand, WH?A+T* does not generate *wut* (because there is no U in the regex), or *wwhat* (because W must appear exactly once), or other words including *wht*, *whatwhat* and *waah*.

In the following problem you will be presented with 15 regexes. To clarify the rule about brackets mentioned above, a string of letters within brackets is considered a single unit. Therefore, for L(OL)+, the first clue, the OL, must be repeated as a single unit. The permissible words for this clue could be LOL, LOLOL, LOLOLOL, and so on, but not LOLO or LOLL.

Mwahahaha!

5

THE WORLD'S FUNNIEST CROSSWORD

Complete the following crossword. The clues are all regex terms, as explained above. The grid contains only letters. Each regex describes a word that is in the grid, so you need to work out two things: the spelling of each word (according to the regex), and which position it should go in.

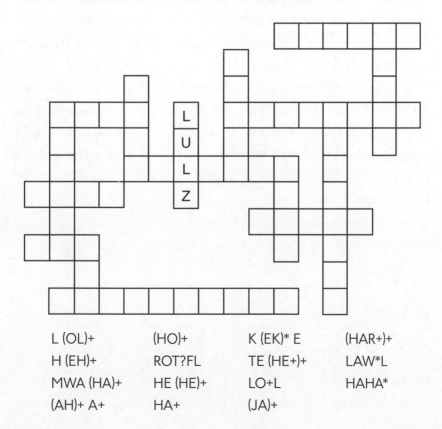

L (OL)+	(HO)+	K (EK)* E	(HAR+)+
H (EH)+	ROT?FL	TE (HE+)+	LAW*L
MWA (HA)+	HE (HE)+	LO+L	HAHA*
(AH)+ A+	HA+	(JA)+	

Another context in which the same word may have many spellings is family names. Smith and Smyth, Brown and Browne, and Hales and Hailes are all pronounced the same but spelled differently. Descendants of the same family often use different spellings of their surname, by either affectation or historical accident.

In 1918, two Americans came up with a system for listing surnames based on how they sound rather than how they're spelled. Soundex is an algorithm—essentially a computer program that existed before the invention of computer—that creates a code for every surname. The idea is that if two names are pronounced the same, their Soundex codes are the same, irrespective of their spellings. The system was used to classify all US census data from 1880 to 1930. (One purpose of Soundex was to file together names that had been misspelled. At a time when many people were functionally illiterate, name misspellings were a frequent occurrence.)

Soundex is still used by genealogists today. If you are looking for relatives that lived in the US a century ago, for example, you can make Soundex searches of US census data. Which is very useful if you don't know whether your great-great-great-granny was a Smith, a Smythe, or a Smithy: All three names have the same Soundex code.

6

WHO DO YOU THINK YOU ARE?

Soundex codes for surnames are created using a step-by-step
process that employs this table:

b f p v	c g j k q s x z	d t	l	m n	r
1	2	3	4	5	6

Here are 16 names and their Soundex codes:

Allaway	A400	Kingscott	K523
Anderson	A536	Lewis	L200
Ashcombe	A251	Littlejohns	L342
Buckingham	B252	Stanmore	S356
Chapman	C155	Stubbs	S312
Colquhoun	C425	Tocher	T260
Evans	E152	Tonks	T520
Fairwright	F623	Whytehead	W330

Once you have worked out the algorithm that produces a
Soundex code, generate the codes for:

Ferguson, Fitzgerald, Hamnett, Keefe, Maxwell, Razey,
Shaw, Upfield

Governments have many lists of their citizens, not just those
provided by censuses. Anyone who has the right to work in
the UK, for example, has a National Insurance number, which
identifies that person in the British social security and tax system.
In Italy, the equivalent is a *Codice Fiscale*, a 16-digit alphanu-
meric expression that encodes your name, gender, date of birth,

and where you come from. In fact, it is possible to deduce all of this personal information from a *Codice Fiscale*.

In the following puzzle, a "Z" indicates that the person in question was born outside Italy. The final letter is the "checksum," a way of making sure that the other elements of the *Codice* are correct. (The checksum is not relevant to the problem, so you can ignore it.)

7

WHAT'S MY NUMBER IN ITALY?

Here are the names of nine people who live in Italy, with their gender, date of birth, country of origin, and *Codice Fiscale*.

Gustavo Aguirre, M, 12/1/80, Argentina	**GRRGTV 80A12 Z600 S**
Veronique Deschamps, F, 16/12/58, France	**DSCVNQ 58T56 Z110 N**
Stefanos Papadopoulos, M, 14/3/50, Greece	**PPDSFN 50C14 Z115 G**
Nalini Sharma, F, 8/6/49, India	**SHRNLN 49H48 Z222 W**
Claudia Torres, F, 10/9/88, Chile	**TRRCLD 88P50 Z603 B**
Miguel Vaca, M, 31/7/68, Bolivia	**VCAMGL 68L31 Z601 R**
Andreas Wackernagel, M, 19/6/76, Switzerland	**WCKNRS 76H19 Z133 G**
Mary Louw, F, 3/5/28, South Africa	**LWOMRY 28E43 Z347 L**
Harry Yeats, M, 9/10/89, USA	**YTSHRY 89R09 Z404 M**

a) What can you deduce about the person with the following *Codice Fiscale*? Give a plausible first name and surname.

SNTPDR 86B03 Z602 Z

b) Elsjana Zogolli was born in Albania on 6/4/82. What are the first 15 symbols of her *Codice Fiscale*?

An important mathematical tool now used by all computer scientists working on natural language problems is a "word embedding," essentially a giant, multidimensional word map created by statistical analysis of a collection of texts. In a word embedding, each word is represented as a point in space. Words that appear in similar contexts in the chosen literature are placed close to one another in the embedding. For example, "cat" and "dog" are words that appear in similar situations, next to words like "pet," "cute," "animal," "vet," "tail," and so on, so the points for "cat" and "dog" will be close. Thus, an embedding is a kind of semantic map of a language based on the company that each word keeps.

Word embeddings usually have about 300 dimensions, although some have more than 1,000. These dimensions have no interpretable linguistic meaning: They are simply the best model for the relative positions of the words. In the following problem, however, the embedding has only two dimensions. (It is not a real embedding. It has been created for the purposes of this problem.) Because words with similar meanings share similar contexts, words with similar meanings will be embedded close together, so be careful to consider all possible meanings of a word.

8

EMBED WITH A LINGUIST

Here are nine words in alphabetical order, and the embedding that maps their semantic relationships:

first	number	second
mathematician	one	time
mathematics	position	two

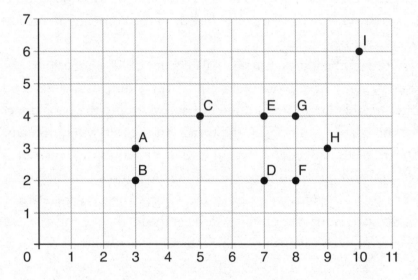

Match the words to their correct positions in the embedding.

Just say that you had a piece of gibberish, or a Dadaist poem, that juxtaposed words that never usually go together. A computer could use an embedding to deduce that this text was not a typical example of natural language. If the words in the text never usually go together they would be far apart in the embedding, raising a red flag that there is something unusual going on. In other words, the embedding provides a pattern the computer can use to make deductions about new material.

To solve the following puzzle, you will have to behave in a similar way. You will be given some snippets of genuine text, from which you will construct a model. You will then be presented with new snippets. By comparing these against your model you will be able to deduce how authentic they are. The puzzle is about "adjective scales," which indicate how different adjectives can express different levels of intensity.

Here's a phrase that makes sense: "good but not great."

And here's one that doesn't: "great but not good."

Something cannot be great without being good because being great assumes being good. ("Great" here taken in the sense of "very good," rather than "important/powerful.") Likewise, it makes sense to say "angry but not furious" but not to say "furious but not angry," because all furious people are necessarily angry. "Good" and "great" express different levels of intensity on one adjective scale, while "angry" and "furious" express different levels of intensity on another. If we use two adjectives from the same scale together, we need to respect their relative positions.

If you were to read, say, an online review of a sprocket that contained the phrases "good but not great" or "angry but not furious," you wouldn't bat an eyelid. But if the review contained the phrases "great but not good," or "furious but not angry," which are obviously wrong, you would (for the purposes of the next problem) deduce that it was written by a bot. If the review contained the phrases "furious but not good" or "great but not angry"—that is, phrases comparing adjectives from *different* scales, which is odd but not obviously wrong—you would be unsure about whether the review was written by a bot, and decide that the situation merits further investigation.

9

A CROMULENT CONUNDRUM

The website ZOINK! is where you go to read reviews of sprockets. The sprocket marketplace is incredibly hip, so everyone writes in the latest slang, which you don't understand. Here are 17 snippets posted on ZOINK! that you know were written by real people, and thus make sense:

1. cromulent but not melaxious
2. not only efrimious but quarmic
3. not only hyxilious but fligranish
4. not only daxic but fligranish
5. not laxaraptic, but just hyxilious
6. not just melaxious but efrimious
7. not only quarmic but nistrotic
8. shtingly, though not efrimious
9. not tamacious, just efrimious
10. not optaxic, just fligranish
11. not only cromulent but shtingly
12. not nistrotic, but just efrimious
13. not nistrotic, just tamacious
14. wilky but not daxic
15. not daxic, just jaronic
16. jaronic but not hyxilious
17. laxaraptic but not optaxic

Based on these examples, can you work out which four of the following snippets were written by real people (since they make sense) and which two were written by bots (since they do not make sense)? You cannot tell whether the three remaining snippets were human- or bot-generated.

A. not only hyxilious but quarmic

B. jaronic but not laxaraptic

C. cromulent but not nistrotic

D. not only tamacious but melaxious

E. not only shtingly but quarmic

F. not fligranish, just wilky

G. optaxic but not hyxilious

H. cromulent but not jaronic

I. not just optaxic but nistrotic

One task that computers do quite well is translation. Thanks to services like Google Translate, it's possible for material in dozens of foreign languages to be translated into English at the click of a button. Computer translation produces its best results with languages that have a large corpus of bilingual material, such as those of the European Union, which publishes all its proceedings in all of its official languages. If a computer has the same text in two languages, it can match words in one language to words in the other, and start to develop a translation algorithm.

The following problem, devised by the computer scientist Kevin Knight in 1997, is a simplified model of bilingual text matching. Simplified, yes, but because languages (even similar ones) use words differently, the matching process is not always straightforward. You are not required to understand the meanings of any of the words. The solution is deduced by comparing where the words appear in the text.

10

WE COME IN PEACE

The year is 2354 CE. Humans have discovered two alien species, one in Alpha Centauri and one near Arcturus. The species speak two different languages, Centauri and Arcturan, neither of which we understand. We have, however, intercepted communications that include a list of twelve Centauri sentences and their Arcturan translations. These are shown below.

	CENTAURI	ARCTURAN
1.	ok-voon ororok sprok	at-voon bichat dat
2.	ok-drubel ok-voon anok plok sprok	at-drubel at-voon pippat rrat dat
3.	erok sprok izok hihok ghirok	totat dat arrat vat hilat
4.	ok-voon anok drok brok jok	at-voon krat pippat sat lat
5.	wiwok farok izok stok	totat jjat quat cat
6.	lalok sprok izok jok stok	wat dat krat quat cat
7.	lalok farok ororok lalok sprok izok enemok	wat jjat bichat wat dat vat eneat
8.	lalok brok anok plok nok	wat lat pippat rrat nnat
9.	wiwok nok izok kantok ok-yurp	totat nnat quat sloat at-yurp
10.	lalok mok nok yorok ghirok clok	wat nnat gat mat bat hilat
11.	lalok nok crrrok hihok yorok zanzanok	wat nnat arrat mat zanzanat
12.	lalok rarok nok izok hihok mok	wat nnat forat arrat vat gat

One day we receive a communication from the Centauri.
We cannot understand it, but we assume it is a message
of peace.

farok crrrok hihok yorok clok kantok ok-yurp

In a gesture of solidarity between humans and aliens, we
decide that this message must be translated into Arcturan
and sent to them.
Your mission is to complete this translation.

Embark on this intergalactic adventure by going through the
Centauri peace message word by word. The message begins with
farok. This word appears in Centauri sentences 5 and 7. The only
word that appears in both those Arcturan sentences, and none of
the others, is *jjat*. So *farok* = *jjat*. (Both *farok* and *jjat* appear in
the same position in sentences 5 and 7, but, as you will discover,
some Arcturan words do not appear in the same positions as their
Centauri translations do.) Continue to *ok-yurp*, and beyond!

Science fiction is all about the present. When you read the
solution to this problem you will realize just how true this is.

Our brief exploration of language and technology ended in outer
space, three centuries into the future. We now return closer to
home, and travel back in time.

LINGO ⒷⒾⓃⒼⓄ

Loan Words

From which languages did English borrow these words?

1. ANORAK
   a) Nepali
   b) Indonesian
   c) Finnish
   d) Greenlandic

2. YO-YO
   a) Bengali
   b) Ilocano (Philippines)
   c) Japanese
   d) Swahili

3. SHAMPOO
   a) Arabic
   b) French
   c) Hindi
   d) Malagasy

4. UKULELE
   a) Hawaiian
   b) Venetian
   c) Guaraní
   d) Portuguese

5. MOSQUITO
   a) Malayalam
   b) Maya
   c) Spanish
   d) Yoruba

6. HOI POLLOI
   a) Fijian
   b) Greek
   c) Russian
   d) Swedish

7. TATTOO (skin art)
   a) Samoan
   b) Sioux
   c) Tupiniquim (Brazil)
   d) Zulu

8. TATTOO (military)
   a) Chinese
   b) Dutch
   c) French
   d) Scots Gaelic

9. TAEKWONDO
   a) Japanese
   b) Turkish
   c) Korean
   d) Thai

10. TOTEM
    a) Maori
    b) Igbo
    c) Rapa Nui
       (Easter Island)
    d) Ojibwa
       (North America)

11. TOMATO
    a) Japanese
    b) Nahuatl (Mexico)
    c) Quechua (Peru)
    d) Urdu

12. COLA
    a) Jamaican Creole
    b) Navajo
    c) Tamil
    d) Temne
       (West Africa)

2
Celts, Counts, and Coats

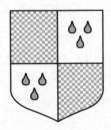

THE LANGUAGES OF THE BRITISH ISLES

The British Isles are an archipelago of exotic indigenous languages and mysterious ancient scripts. Its inhabitants speak almost a dozen native tongues (and at least two indigenous sign languages). You are presumably fluent in at least one, the evolution of which will be the theme of this chapter. First, though, let's journey back in time to the Early Middle Ages, where we will discover an ancient alphabet found nowhere else on Earth.

Ogham—in Irish pronounced "ome," to rhyme with "home"—was a script used in Ireland between the fifth and ninth centuries. It survives mainly in the form of 330 or so inscriptions on stone monuments, which are the earliest written examples of the Irish language. The alphabet is of huge interest to linguists and historians, and nowadays is a proud symbol of Irish identity (as the tattoo parlors of Dublin and Cork will attest). Ogham words consist of a vertical stemline with notches or strokes cut across it, like a child's drawing of a fish bone. Each group of notches or strokes represents an individual letter.

11

OGHAM, SWEET OGHAM

Here are some names of plants in Old Irish written in both
Ogham and the Latin alphabet. The English translations are
for information only, and do not contribute to the solution of
the problem.

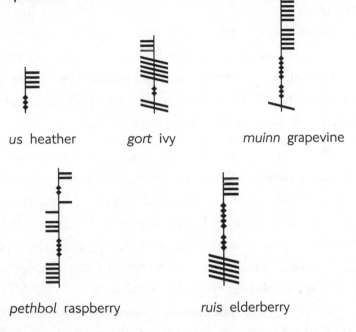

us heather   *gort* ivy   *muinn* grapevine

pethbol raspberry   *ruis* elderberry

Write in the Latin alphabet:

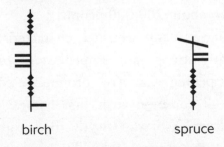

birch   spruce

Write in Ogham:

nim ash   *luis* elm

The most recent Ogham stone inscription dates from the seventh century, although scribes must have had a working knowledge of the alphabet for a few hundred years after that, since historians have also found a few words in ink. The most curious example, in the margins of a ninth-century Old Irish manuscript (that uses the Latin alphabet), is an Ogham doodle that reads "massive hangover," presumably the scribe's secret confession. Some things never change.

Divination using Ogham is now popular among British pagans, who are attracted to the idea that it is an authentic, pre-Christian cultural artifact—although it probably isn't. Ogham was most likely inspired by the Latin alphabet, as used by Christian Romans in the early centuries CE.

When the Romans first came to Britain, in 55 BCE, the locals all spoke Celtic languages. The Celts were a European people who, during the previous 1,000 years, had spread over much of the continent, from Orkney to Turkey and from Portugal to Poland. The Celtic languages became extinct in mainland Europe by around the sixth century CE, but survived in the westernmost fringes of the British Isles. Together, the big three—Welsh, Irish, and Scots Gaelic—are spoken today by about a million people. (In the Early Middle Ages, some Celtic language speakers emigrated from Britain to Brittany. Their language evolved into Breton, now spoken by about 200,000 people.)

Celtic languages have many unusual characteristics that make them particularly strange and impenetrable to outsiders. First of all, the spelling. Can you pronounce Meadhbh, Caoimhe or Tadhg, all common Irish first names? Or how about the Welsh villages Ysbyty Ystwyth, Bwlch-gwyn and Llanfairpwllgwyngyllgogerychwyrndrobwllllantysiliogogogoch? (Answers at the bottom of page 38.)

"Mutation," another exotic and distinctive feature of Celtic languages, is the propensity of the initial consonant of a word to change depending on how that word is used in a sentence. In Welsh, for example, the city of Bangor is usually written *Bangor*. Were you to say, however, that you went *to* Bangor, then you would say you went to *Fangor* (pronounced "vangor," since the "f" in Welsh is pronounced "v"). On the other hand, if you were *in* Bangor, you'd say you were in *Mangor*.

Bangor, Fangor, Mangor? Let's call the whole thing off.

The change of a "b" to an "f" is known as a "soft mutation." The consonants "c," "d," "m," and "t" also have soft mutations, as you are about to discover. There are at least 30 grammatical rules concerning when a Welsh consonant undergoes a soft mutation. Thankfully for you, the next problem concerns only three of them.

12

IN THE STREET, HE SAW A MUTANT!

Here are some Welsh nouns in their unmutated forms with their English translations:

beic	bicycle	*dafad*	sheep
ceffyl	horse	*darlun*	picture
ci	dog	*theatr*	theater
cath	cat	*bachgen*	boy
tad	father		

Here are some Welsh sentences with their English translations. The order of words in Welsh is important. The *dd* is pronounced "th," as in "this," and *ll* is pronounced by putting your tongue on the roof of your mouth as if you are about to say an "l," and blowing.

Aeth Megan i Fangor	Megan went to Bangor
Aeth Emrys i Aberystwyth	Emrys went to Aberystwyth
Mae dafad yma	A sheep is here
Mae yma ddafad	Here is a sheep
Mae yn Aberystwyth dad	In Aberystwyth is a father!
Mae Megan yn Llangollen	Megan is in Llangollen
Gwelodd Megan fachgen	Megan saw a boy
Gwelodd ddarlun	She saw a picture
Gwelodd y dyn gath	The man saw a cat

Answers from page 36: "Maeve," "Queevah," and "Tyge." The Welsh villages are pronounced "Uss-butty Uss-twith," "Bull-ch-gwin," and . . . don't even think about it. (In fact, the long name is a nineteenth-century publicity stunt that succeeded beyond all expectations. The village's official name is Llanfair Pwllgwngyll, even though the long version is used extensively for tourism. If the long version were to be used "properly," however, it would contain many hyphens, separating the constituent words.)

Choose the correct Welsh translation for each English sentence below.

1. He saw a bicycle in the street
   a) *Gwelodd beic yn y stryd*
   b) *Gwelodd feic yn y stryd*
   c) *Gwelodd yn y stryd beic*
   d) *Gwelodd yn y stryd feic*

2. In the street, he saw a bicycle!
   a) *Gwelodd yn y stryd feic!*
   b) *Gwelodd yn y stryd beic!*
   c) *Gwelodd beic yn y stryd!*
   d) *Gwelodd feic yn y stryd!*

3. In the theater, she saw a horse!
   a) *Gwelodd yn y theatr geffyl!*
   b) *Gwelodd ceffyl yn y theatr!*
   c) *Gwelodd yn y theatr ceffyl!*
   d) *Gwelodd geffyl yn y theatr!*

4. The boy's father saw a dog
   a) *Gwelodd dad y bachgen gi*
   b) *Gwelodd tad y bachgen gi*
   c) *Gwelodd tad y bachgen ci*
   d) *Gwelodd dad y bachgen ci*

Linguistic mutations arise because as languages change and evolve, so do their pronunciations. Depending on factors such as the preceding sound, "t"s become "d"s, or "b"s become "f"s. And so on. You can hear this happening already in English: An American saying the word "metal" would pronounce it closer to "medal." The consonant has softened from a "t" to a "d" just like it does in a Welsh mutation. In the Celtic languages, however, the spellings were altered to reflect changes in pronunciation, something that has not yet happened in American English. Goddit? Whaddever.

Now that you're an expert on these grammatical griffins, prepare to meet even more terrifying mutants from across the Irish Sea.

Ireland is one of the most densely named countries in Europe. The number of named administrative units it has—which are divided into counties, baronies, parishes, and townlands—is around 65,000. Were you to drive around Ireland, you would notice that all these places have two names: an Irish one and an English one. Since 2012 both names have equal legal status, the culmination of a national commission set up in the 1940s to restore the original Irish names after centuries of Anglicization. (A standardized English-language spelling was introduced when the Ordnance Survey did its first large-scale survey of Ireland in the nineteenth century.)

Many Irish places are named after geographical features. In some cases, the English name is a direct translation of the Irish one. For example, Highpark is An Pháirc Ard, since *páirc* means "park," and *ard* means "high." In other cases, the English name is a phonetic imitation of the Irish one. For example, Ballynaparka is how you pronounce *Baile na Páirce*, which means "Town of the Park." Did you see what just happened? The Irish word for "park" can be written as *páirc*, *pháirc*, and *páirce*. The following problem, which requires you to deduce which word form is used in which situation, is the most fiendish in the book so far.

13

TOPONYM O' THE MORNING

Here are some Irish place-names. The first twelve have
English names that are phonetic imitations of the Irish names.
The remaining five have English names that are translations.

English	Irish	Translation
Ballynaparka	*Baile na Páirce*	Town of the Park
Binbane	*An Bhinn Bhán*	The White Peak
Bunagortbaun	*Bun an Ghoirt Bháin*	Base of the White Field
Buncurry	*Bun an Churraigh*	Base of the Marsh
Clonamully	*Cluain an Mhullaigh*	Meadow of the Summit
Curraghmore	*An Currach Mór*	The Big Marsh
Dunard	*An Dún Ard*	The High Fort
Glenamuckaduff	*Gleann na Muice Duibhe*	Valley of the Black Pig
Gortnakilly	*Gort na Cille*	Field of the Church
Kilcarn	*Cill an Chairn*	Church of the Mound
Kilknock	*Coill an Chnoic*	Wood of the Hill
Killeshil	*An Choill Íseal*	The Low Wood
Blackabbey	*An Mhainistir Dhubh*	The Black Abbey
Highpark	*An Pháirc Ard*	The High Park
Castlepark	*Páirc an Chaisleáin*	Park of the Castle
Whitefield	*An Gort Bán*	The White Field
Woodland	*Talamh na Coille*	Land of the Wood

**What are the Irish names and English translations of these
towns and villages?**

1. Mullaghbane
2. Knocknakillardy
3. Gortnabinna
4. Blackcastle

This is a very tricky problem, not to be done at the wheel.

I'll set you up. First, draw up a vocab list. You know the Irish for "park," "town," and "high." By comparing patterns among the Irish names and the English translations you can deduce the meanings of the rest of the words. That should give you a good idea of the English translations of the four places in the question. What's much harder is working out the correct way to spell each of the Irish words.

When you look at the list of place names, you'll notice that they have only two types of structure. They might be made up of one noun and one adjective, in which case the Irish name begins with *An*, meaning "the." Or they might consist of two nouns (and maybe an adjective) in the form "X of the Y." In that case the Irish name has the word *an* or *na*, meaning "of the," in the middle of it. Written more formally, these two types of structure are as follows:

i) *An* [noun] [adjective]

ii) [noun] *an/na* [noun] [optional adjective]

There are thus three different positions a noun can take in an Irish place-name: the beginning of the name (when it is one of two nouns); following *An* at the beginning of the name (when it is the only noun); or preceded by *an* or *na* when it is the second noun. Let's call these positions 1, 2, and 3. Draw up a table of all the nouns in the list above, sorting them into these three positions.

If you are a genius at languages, you will have done this right away, and you will be well on your way to solving the problem. For most of us, however, even with this table to analyze, the pattern is still quite hard to decipher, since it relies on a tiny detail that is obscured in a noise of unfamiliar Celtic spellings.

The detail is the letter "i." Irish nouns can be divided into those whose final vowel (of the basic form) is an "i," and those whose final vowel isn't. (Position 1 presents the basic form.) *Páirc* falls into the first category. *Gort* falls into the second. Both classes of noun mutate in different ways depending on which position they are in. As do the adjectives that descibe them.

You now have enough information to make a stab at the answers. Note that Knocknakillardy has two possible Irish names because the English pronunciation "kill"/"kil" can refer to any of *Cill*, *Cille*, *Coill*, *Choill*, or *Coille*. I hope you kill/kil this problem, before it kill/kils you.

The Germanic tribes that landed on Britain's eastern shores in the fifth century CE conquered most of the island, pushing the Celtic-language speakers to its northern and western extremities. These immigrants, the Anglo-Saxons, were the first speakers of a new indigenous tongue, English. Although to the ears—and eyes—of English speakers nowadays, the language they spoke is almost as unintelligible as Irish or Welsh.

14

WE ALL LOVED THE GIRL

Here are some sentences in Old English and their translations in Modern English. The letter þ is pronounced like the "th" in "thin"; the letter æ like the "a" in "cat."

wit lufodon þæt mægden	we two loved the girl
þæt mægden unc lufode	the girl loved us two
ge lufodon þone cyning	you all loved the king
se cyning inc lufode	the king loved you two
þæt mægden we lufodon	we all loved the girl
we inc lufodon	we all loved you two
wit eow lufodon	we two loved you all
unc lufode se æþeling	the prince loved us two
þæt cild ge lufodon	you all loved the child

Translate into modern English:

1. se cyning eow lufode
2. ge lufodon þæt mægden
3. wit inc lufodon

Translate into Old English:

4. The prince loved the child
5. The child loved the prince
6. We all loved the child
7. The child loved you two

Once upon a time in Germany there lived a clever boy called Jacob Grimm. Jacob loved words and languages. In particular, he loved to study the connections between the same words in different languages.

"Isn't it interesting," he thought to himself as he ate his perfectly cooked porridge, "that in German, English, and Icelandic, words with similar meanings often look very similar?"

"And how curious it is," he reasoned as he nibbled on a gingerbread man, "that certain sounds in one language usually correspond to the same sounds in another. For example, a "d" in a German word is often a "th" in an English word, such as *der*/the, *drei*/three, *danke*/thanks.

"I know what I will do," exclaimed Jacob one morning as he crunched on a red apple. "I will write a big long book with all my ideas in it!"

Jacob published his book, *Deutsche Grammatik*, in 1822, and it brought him huge acclaim—at least among linguists—because of its scientific approach to analyzing how pronunciation changes over time. Before his book, scholars had thought that word sounds evolved haphazardly. But Jacob showed that sound shifts transform all the words of a language systematically. For example, it wasn't just that, say, some "p"s become "f"s when one language evolves into another—*all* "p"s become "f"s. Jacob's theory that changes affecting consonants like "p" and "f" were regular and exceptionless across a group of languages is known as Grimm's law. It revolutionized historical linguistics, the discipline that aims to reconstruct ancestral languages by studying patterns of sound change.

And Jacob lived happily ever after.

Grimm, of course, is much better known as the compiler (together with his younger brother Wilhelm) of a book of fairy tales. In fact, the brothers' interest in folk stories stemmed from their desire to study, and preserve, the German language. A century later, J. R. R. Tolkien's fascination with the origins of English became the foundation for his fantasy novels. It is notable that Grimm and Tolkien, the creators of two of the best-known fairytale universes, both started their careers as historical linguists.

The next puzzle will explore the magical world that Grimm brought to life. It involves the concept of a "proto-language," a language hypothesized to have existed, from which other known languages evolved. The ancestor of English, German, and the Scandinavian languages is called Proto-Germanic.

Historical linguists believe Proto-Germanic was spoken 2,000 years ago in the area around Denmark, southern Norway, and southern Sweden. (By convention, a word in a reconstructed proto-language is always preceded by an asterisk, as in *krampaz*, to indicate that it is a hypothesized form.)

15

IT'S GRIMM UP NORTHWEST EUROPE

Fill in the blanks in the following grid. The letter ð, or "eth," indicates a voiced "th" as in "this," the letter þ, or "thorn," indicates a voiceless "th" as in "thin," and the æ sounds like the vowel in "bed." The (¯) in Proto-Germanic, and the (') in Icelandic both indicate long vowels. The letter *j* in Proto-Germanic and German is pronounced "y." Nouns in German are capitalized. The letter *w* in German is pronounced "v."

Proto-Germanic	English	German	Icelandic
*krampaz	cramp	Krampf	krampar
*aplu	_ _ _ _ _	_ _ _ _ _	epli
*swanaz	swan	Schwan	svanur
*þrīz	three	drei	þrír
*swīnan		_ _ _ _ _	_ _ _ _ _	_ _ _ _
*jæran	year	Jahr	ár
*þūman	_ _ _ _ _	Daumen	þumalfingur
*þurnuz	_ _ _ _ _	_ _ _ _	þyrnir
*wurðan	_ _ _ _	Wort	orð
*_ _ _ _ _ _ _ _	sword	Schwert	sverð

Because sound shifts are regular between proto-languages and their descendants, you can fill in the grid much as you would a Sudoku puzzle, using deductive logic across both rows and columns. Indeed, this puzzle is a simplified form of the kind of work historical linguists do every day. (You may have noticed that Icelandic seems to have changed the least since Proto-Germanic, a consequence of Iceland's geographic isolation.)

Other languages descended from Proto-Germanic include Afrikaans, Faroese, Scots, and Yiddish. And Dutch, the closest major language to English, so I'm going to sneak it into this chapter.

Despite its similarities to English, spoken Dutch is notoriously hard to make sense of. Hence the phrase "double Dutch": something twice as unintelligible as the already incomprehensible. The next problem involves a different type of double Dutch: the language's unusually large number of homonyms, or words that sound or look similar but have multiple meanings, and which are the basis of many puns. For example, the word *deken* means "deacon," *laken* means "to strongly disapprove of," and *kussen* means "kissing/to kiss." So the sentence *Dekens laken kussen* means "Deacons strongly disapprove of kissing." However, *deken* also means "blanket," *laken* also means "bedsheet," and *kussen* also means "pillow." Thus, the sentence can also be read: "Blankets bedsheet pillow."

For more hilarious homonyms from Holland, read on.

16

TRIPLE DUTCH

The following sentences are grammatically correct. Their translations are listed below in random order.

The words *vliegen* and *weg* each have two meanings. What are they? What are the three meanings each of *bij* and *graven*?

1. *Wij laken het graven graven.*
2. *Graven graven nooit graven.*
3. *Als vliegen vliegen laken, vliegen vliegen nooit.*
4. *Als achter vliegen vliegen vliegen, vliegen vliegen vliegensvlug.*
5. *Twee vliegen vliegen over de weg weg en er is een bij bij.*
6. *De weg is weg.*
7. *De vliegen vliegen bij de bij.*

a. The road is away.
b. If flies fly behind flies, flies fly as fast-as-a-fly.
c. If flies strongly disapprove of flying, flies will never fly.
d. The flies fly near to the bee.
e. Counts (noblemen) never dig graves.
f. Two flies fly over the road away and there is a bee with [them].
g. We strongly disapprove of the digging of graves.

Flying fast-as-a-fly back to Britain, we resume the story of English. In 1066, the Normans invaded, which began a period during which the ruling elite spoke Anglo-Norman, a dialect of Old French. This was a time of jousting knights, chivalric codes of honor, and patterned armorial bearings. In 1484, Richard III set up the College of Arms, the official heraldic authority in the kingdom, whose job it was to grant new coats of arms.

When the College of Arms grants a new coat of arms, the official documentation contains a formal description of the pattern in "blazon" (the "a" is pronounced as in "black"), the technical language of heraldry. Blazon has its own syntax, grammar, and vocabulary, mostly derived from Old French. For administrative purposes, the blazon description is what's important about a shield: It defines the pattern precisely and succinctly, enabling the College of Arms to make sure each coat of arms is unique. It also means that a person can easily communicate what's on the shield with no ambiguities about the color, symbols, or arrangement. A blazon description from 1484 is as clear as is one from 2021.

The College of Arms currently grants around 150 coats of arms a year, all of them with official blazon descriptions. To get one will cost you about £7,000 (almost $10,000). It's a lot of money, although they do last forever.

17

AN ARMOIRE OF COATS

Match the nine shields shown below to their correct blazon descriptions opposite. Use the color key to determine the colors in the shields (or maybe break out your crayons)!

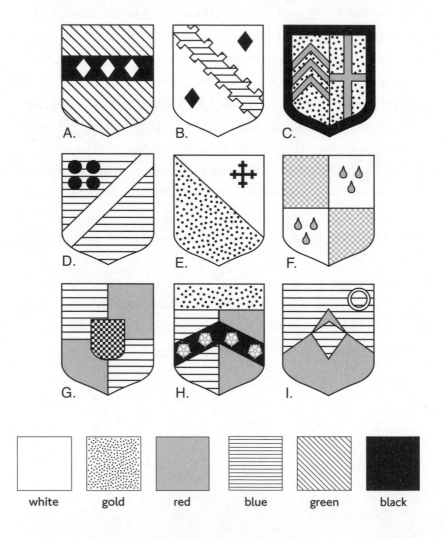

1. *Quarterly 1 & 4 chequy Gules and Argent 2 & 3 Argent three gouttes Gules two one*
2. *Azure a bend sinister Argent in dexter chief four roundels Sable*
3. *Per pale Azure and Gules on a chevron Sable four roses Argent a chief Or*
4. *Per chevron Azure and Gules overall a lozenge counterchanged in sinister chief an annulet Argent*
5. *Quarterly Azure and Gules overall an escutcheon chequy Sable and Argent*
6. *Vert on a fess Sable three lozenges Argent*
7. *Argent a bend embattled between two lozenges Sable*
8. *Per bend Or and Argent in sinister chief a cross crosslet Sable*
9. *Or three chevrons Gules impaling or a cross Gules on a bordure Sable*

In the blank shields below, draw the patterns described by the following blazons:

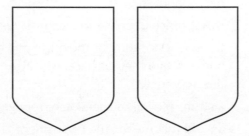

10. *Quarterly Azure and Or overall five lozenges Sable three one one a bordure Vert*
11. *Per chevron Argent and Gules in sinister chief a roundel Or in dexter chief a cross crosslet Sable*

In the centuries after the Norman conquest, the English language incorporated many words from Old French. A few centuries after that, English began to plunder an even older European language, Ancient Greek. It has long been the custom that scientists and inventors, upon discovering or building something new, reach for their Classical dictionaries. For example, when the Victorian paleontologist Richard Owen wanted a name for the giant reptile-like skeletons found in fossils, he borrowed the ancient Greek words *deinos* and *sauros*, for "terrible" and "lizard," thus coining the term "dinosaur." For the following problem I am assuming you are familiar with some basic scientific terms.

IT'S ALL GREEK TO ME

Here are some English words with Ancient Greek roots.

Agoraphobia	an abnormal fear of open or public places
Alexander	defender of men
Anesthesia	the absence of sensation, especially pain
Anthropology	the scientific study of the origin and behavior of humans, and their physical, social, and cultural development
Antipathy	a strong feeling of aversion or repugnance
Antonym	a word that means the opposite of another word
Bibliophile	a lover/collector of books
Dactylonomy	the art of counting with fingers
Demagogue	a leader who appeals to popular prejudices
Dendroclastic	a destroyer of trees
Epeolatry	the worship of words
Fibromyalgia	muscle pain
Gastrosoph	a person skilled in matters of eating

Hippodrome	an arena for horse races
Misogyny	a hatred of women
Morology	the study of fools
Pediatrics	branch of medicine relating to children
Pelophilous	mud or clay-loving
Polygamy	the condition or practice of having more than one spouse at one time
Tachycardia	a rapid heart rate
Telepathy	communication by means other than the senses.

Derive the meanings of the following English words, several of which are very rare:

Bibliophobia	Misandry	Mystagogue
Cardialgia	Misanthropy	Pelotherapy
Dendrolatry	Misogamy	Philanthropism
Dromomania	Misopedia	Photagogue
Gynophilia	Monandry	Polydactyl
Hippophobia	Morosoph	Telesthesia

The large number of words English has incorporated from other languages has given it a rich and vibrant vocabulary, at the expense of turning it into an orthographic omnishambles. Words of French origin tend to be spelled using French spelling rules, words of Greek origin using Greek spelling rules, and so on, meaning that one letter can be pronounced in myriad ways. Of the many attempts at spelling reform in the UK, none has had any success, even with vocal support over the centuries from public intellectuals such as John Milton, Samuel Johnson, Charles Dickens, and Charles Darwin.

In the twentieth century, the most famous crusader for sensible spelling was the playwright George Bernard Shaw. When he died in 1950, he left money in his will for the invention and promotion of a phonetic alphabet, stipulating that the letters should be as distinct from the Latin alphabet as possible, so readers didn't think that the new script was just bad spelling. Within a few years of his death, a competition for a new alphabet had been launched. One of the four winners, Ronald Kingsley Read (a beautiful example of nominative determinism), amalgamated their entries and named the collective invention the "Shavian alphabet." Almost all the money allocated to spelling reform in Shaw's will paid for Penguin to publish, in 1962, an edition of his play *Androcles and the Lion*, in both the Shavian and Latin alphabets. It is unclear, however, if the Shavian section was ever read.

19

SURE, SURE, MR. SHAW

The list on the left has five phrases written in the Shavian alphabet. The transliterations on the right are listed in random order. Match the phrases to their correct transliterations.

1. 𐑞 𐑒𐑨𐑑 𐑕𐑤𐑧𐑐𐑑	a. this is Shavian
2. 𐑢𐑰 𐑣𐑨𐑝 𐑒𐑨𐑑𐑕	b. the cat slept
3. 𐑓 𐑧𐑝𐑼	c. to learn
4. 𐑞𐑦𐑕 𐑦𐑟 ·𐑖𐑱𐑝𐑾𐑯	d. we have cats
5. 𐑑 𐑤𐑻𐑯	e. for ever

Some Shavian characters come in pairs, such as 𐑕 and 𐑟, 𐑓 and 𐑝.

By thinking about the sounds these characters make, what is the character for "b"?

One reason spelling reform has never had any traction (apart from in a very minor way in the US, where some spellings have been simplified) is that inconsistent orthography has hardly held English back in its remarkable rise to become the most spoken language in the world.

In terms of native speakers, English is only third in the list of most-spoken languages, after Mandarin Chinese and Spanish. But once you include people who speak it as a second language, English surges ahead of the competition, with almost 2.3 billion speakers. Indeed, about two out of every three English speakers speak it as a second language. The global role of English, overall, is as a lingua franca that enables people who do not speak the same language to communicate with each other.

One place that has embraced English as an auxiliary language is Papua New Guinea, the country in the southwest Pacific that covers half of the island of New Guinea and its offshore islands. Papua New Guinea is the most linguistically diverse place on Earth. Its population of eight million speak about 800 native languages. (On average, that's a different language for every 10,000 people, although the number of speakers of each varies, from a few dozen to a few hundred thousand.) Quite why Papua New Guinea has so many languages is still unexplained. One reason may be geographical; many communities are isolated even from their closest neighbors due to mountains, jungle, or swamp, thus allowing individual languages to develop free from outside influences. Current opinion, however, suggests a more important factor is one of social organization: the fissiparous nature of Papuan society has created a situation where neighboring communities—even those not separated by geographical obstacles—determinedly maintain a clear social distance.

Papuan languages are, on the whole, mutually unintelligible. So when two Papuans from different communities want to converse with one another, they will often speak Tok Pisin, a creole made up mostly of English words spoken with Melanesian pronunciation and syntax. Tok Pisin started off in the nineteenth century as a pidgin—a hodgepodge of borrowed words with barely any grammatical rules—but it has gradually evolved into an official, living language with native speakers, and is now the most widely spoken language in the country.

Many words in Tok Pisin are instantly recognizable to English speakers, yet the concepts expressed by the words can be very different. For example, the word *botol* comes from "bottle." But if you asked a Papuan for an empty bottle using *botol* they would not understand you, since *botol* actually means a full bottle. If you want the container, you need to ask for the *kontena*. "The fascinating thing about Tok Pisin is that it is the easiest language to speak for an English native speaker, but you really need to be a philosopher to understand what you're talking about," says the French philosopher, and fluent Tok Pisin speaker Maxime Rovère.

20

WARI BILONG YU

Here are some words and phrases in Tok Pisin:

haus	house
haus bilong mi	my house
haus moni	bank
haus sik	hospital
maus	mouth
lek bilong pik	leg of pig, ham
wari bilong yu	your problem
gras bilong dok	fur of dog
gras nogut	weeds
meri	woman, wife, girl
wara	water, river
solmit	salted meat
susok	footwear
man bilong toktok	chatterbox

Translate into Tok Pisin: leg of dog, saltwater, grass, bed

Translate into English:

haus bilong yu	gras bilong het	klos meri
haus bilong king	gras bilong fes	tekewe klos
haus bilong wasim	maus gras	bret
klos	katim gras (two	kukim bret
haus dok sik	meanings)	kikbal
haiskul	pen bilong maus	susok man

———◆———

Our attention now turns from English to another type of international vernacular: the "universal" language of numbers.

LINGO Ⓑ Ⓘ Ⓝ Ⓖ Ⓞ

Polish Phonetic Spelling

In Poland in 1957, at the height of the Cold War, a map of Europe was printed in which all names were spelled according to Polish pronunciation rules. The 14 towns below, which include Ashford, Colchester, Gillingham, Herne Bay, Margate, Newhaven, Upminster, and Whitstable, appeared in the section that covered the southeast of England.

   Can you match these words to their Polish spellings and deduce the English spellings of the remaining towns?

1. *Apmynste*

2. *Datfed*

3. *Douwe*

4. *Dzylynem*

5. *Eszfed*

6. *Hejstynz*

7. *Hen-Bei*

8. *Istbon*

9. *Koulczyste*

10. *Luys*

11. *Łytstebl*

12. *Magyt*

13. *Njuhejwn*

14. *Saufend-on-Sji*

3

All About
That Base

WORDS THAT COUNT

The world's oldest surviving written documents are bureaucratic gobbets that date from Sumer (now Iraq) from the late fourth millennium BCE. Whoever wrote these "texts" did so by pressing a reed into a palm-sized clump of wet clay and leaving it to dry in the sun. Sumerian scribes drew small pictures of objects, such as cows or donkeys, each one next to symbols representing numbers. The resulting text was the ancient version of an Excel spreadsheet, giving administrators an efficient way to keep track of stock. Sumerian clay tablets mark the beginning of the history of writing, and demonstrate that literature only got going thanks to the human desire to count.

In this chapter we'll explore the language of numbers. Our journey will encompass the history of numerical notation from Mesopotamia to the modern world, and we will learn to count from one to ten in Maori, Malagasy, and Manx. Numbers are very basic concepts—yet as we shall see, speakers of different languages count in very different ways.

Our first stop is New Caledonia in the South Pacific, where the indigenous language Iaai is spoken by some 4,000 people.

21

AYE, IAAI

The following list contains seven numbers written in Iaai. It is a sequence in which each term adds 2 to the one before. (So, for example, the numbers could be 2, 4, 6 . . . or 11, 13, 15 . . .)

thabung ke nua lo

thabung ke nua vak

libenyita ke nua khasa

libenyita ke nua kun

libenyita ke nua thabung

libenyita ke nua thabung ke nua lo

libenyita ke nua thabung ke nua vak

What are these numbers?

It's not just little-spoken languages like Iaai that contain number words that are very different from our own. Danish, with six million native speakers, has a counting system that, to nonspeakers, is baffling at first sight.

22

THE KNIGHTS WHO SAY NI

Here are some numbers in Danish:

fire	4
enogfirs	81
seksogtres	66
toogtyve	22
fem	5
syvoghalvtreds	57
ni	9
nioghalvfjerds	79
tre	3

What are the following numbers?

seks

nioghalvtreds

treogtyve

femoghalvfems

toogtres

halvfjerds

What are these numbers in Danish?

7

21

54

85

99

Now back to the tablets.

By around 2000 BCE, the primitive number symbols of Sumer had evolved into a fully functional notation that could describe, clearly and efficiently, numbers from 1 up to the tens of thousands. The style of writing in which a reed is pressed into clay is known as "cuneiform" (pronounced cue-NAY-ih-form). Cuneiform numerals comprised two types of mark: a thin vertical one, made with the flat end of the reed stylus, and a thicker horizontal one, made with the corner edge. This system, used in ancient Babylonia, was arguably the most mathematically effective of the ancient world, certainly until the first centuries CE. It permitted Babylonians to describe very big numbers, and (using their equivalent of the decimal point) very small subdivisions of numbers. This enabled them to observe and document movements of the night sky. Babylonian astronomy is the foundation of Western astronomy, and the reason our clock faces look the way they do.

23

MATH BABYLON

Here are some Babylonian numbers:

𒐏	40
𒐗 𒌋𒌋 𒐗	386
𒐕 𒌋𒐕	71
𒐖 𒌋 𒐖	132
𒐖	62
𒌋 𒐏 𒐖	654
𒐏 𒐗	59

Which numbers do the following symbols represent?

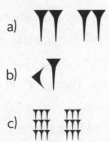

a) 𒐖 𒐖

b) 𒌋𒐕

c) 𒐗 𒐗

The Inca Empire was the largest in the Americas before the arrival of the European conquistadors. Unlike the Bronze Age civilizations of the Middle East, the Incas had no known system of writing, although they did invent their own method for recording numbers that used knots tied on strings. A *khipu* is a collection of these knotted strings, tied together in a bundle. *Khipus* can be made up of hundreds of strings, often in many colors, and they were used to record dates, taxes, census information, and measurements across the Inca Empire.

Archaeologists studying the pre-Columbian Americas knew that *khipus* recorded numbers, but they did not know exactly how, until Leslie Leland Locke, a high school math teacher in Brooklyn, deciphered the system in 1912. In the next problem you will replicate his decipherment by looking at a segment of the *khipu* he used to make his breakthrough. (It is now an exhibit in New York's American Museum of Natural History.) The "primary cord" does not convey any numerical information: its purpose is simply to join all the other strings together.

24

A LOAD OF OLD ROPE

The image opposite shows a section of a *khipu* laid out flat. The horizontal line is the primary cord. Each string attached to the cord represents a three-digit number. Each set of four strings is grouped together by a fifth string. The symbols "x" and "o" denote two different types of knot.

The image overleaf shows another set of four strings attached to the same *khipu*. Again, they are grouped together by a fifth string, which I have left blank.

Fill in the knots on the empty string, marked with a "?".

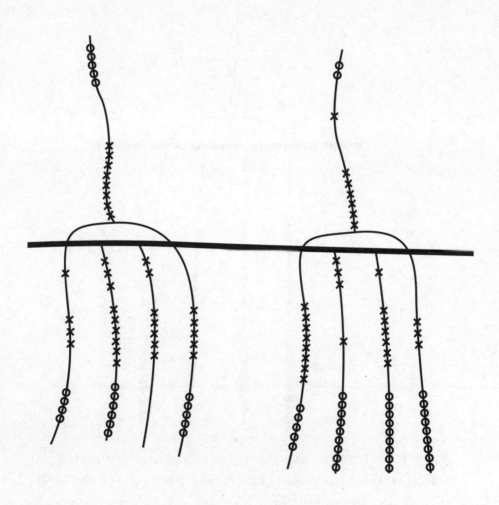

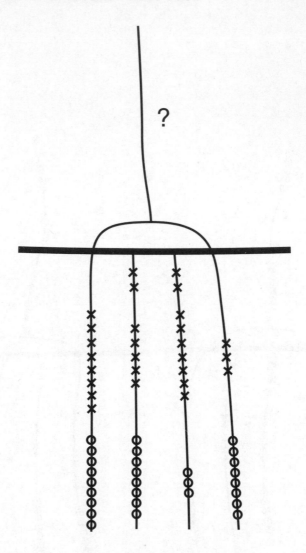

Count the knots in each image and speculate how each string might encode a three-digit number. Remember, the system is very simple. Do not overthink it. Once you have a theory for how the numbers are encoded by the knots, look at each set of five strings separately. Each set of five strings encodes five three-digit numbers. In what way are these five numbers related? Once you find that out, a piece of basic arithmetic will lead you to the answer.

About a thousand *khipus* survive. In recent years, historians have started to think that some, which are unusually complex in terms of their color, ply direction, and material, may, in fact, be examples of a "writing system" in which the knots somehow denote phonetic sounds. The words that may be hidden in these Incan strings are a tantalizing mystery awaiting the next generation of decipherers.

In ancient Europe, the Greeks and the Romans both plundered their alphabets for numerical notation. The Greeks used 27 letters for 27 different numbers: the digits from 1 to 9, the tens from 10 to 90, and the hundreds from 100 to 900. The Roman system was simpler: seven letters, I, V, X, L, C, D, and M, which stood for 1, 5, 10, 50, 100, 500, and 1,000.

Extra puzzle: What is the biggest number written in Roman numerals that is also a word in English? (The answer is in the back.)

Roman numerals would ultimately be replaced by the decimal notation in use today. This notation arrived in Europe in the thirteenth century, brought by traders sailing to and from Arab north Africa, which is why the symbols are commonly known as "Arabic" numerals. It was superior to all previous systems because it included a symbol for zero, which made arithmetic easier. The description of these numerals as Arabic, however, is misleading. The notation originated in India, around the fifth century CE.

It is to Classical India we now turn. The scientists of that period used several types of notation, not just the precursor to our Arabic numerals. One of these notations was invented by the first of the major Indian astronomer-mathematicians, Aryabhata (476–550), and was used only by him. Like Greek and Roman numerals, it was based on an alphabet, although Aryabhata's system was incomparably better: it enabled him to write every whole number from 1 to 1,000,000,000,000,000,000,000 in a brilliantly concise way. I've decided to include a puzzle about this system because I love its ingenuity, and also because it's curious that the greatest mathematician of his age preferred it to the much simpler "decimal-with-zero" system that would eventually become the world standard (and which he was aware of). Via Aryabhata's numbers, I can also introduce the Devanagari alphabet, the world's fourth most-used script (after Latin, Chinese, and Arabic)—just in case you think I'm diverging too much from this book's subject of language and languages!

Devanagari was used for Sanskrit (the language Aryabhata wrote in) and is used today in more than 100 other Indian languages, including Hindi. Devanagari is actually an "abugida" or "alphasyllabary," a halfway house between an alphabet, in which there is a symbol for every sound; and a syllabary, in which there is a symbol for every syllable. In Devanagari, each consonant has its own symbol, which, when presented alone, is pronounced with the vowel "a." So, for example:

प *pa*　　क *ka*　　द *da*

To get the other vowel sounds, you add a diacritic, the name given to subsidiary squiggles like accents or cedillas in the Latin alphabet. For example, the diacritic for "i" is ि and for "u" it's ु. Thus:

पि *pi*　　कि *ki*　　दि *di*
पु *pu*　　कु *ku*　　दु *du*

(Note that Devanagari reads left to right, and the diacritic for "i" precedes the consonant even though it is pronounced after it.)

In Aryabhata's number notation, the basic elements are syllables, not consonants or vowels. Each syllable represents a number, and each number can be written using a sequence of one or more syllables. Note that the consonants *kh*-, *gh*-, and *dh*- are the aspirated versions of *k*-, *g*-, and *d*-, which means that they are pronounced with an extra burst of breath.

25

NANU NANU

Below is a list of numbers written in Aryabhata's numerical notation, transcribed from Devanagari into the Latin alphabet.

1	ka	100	ki	10,103	gakiku
2	kha	101	kaki	30,101	kakigu
3	ga	218	dakhi	160,423	baghitu
18	da	1,923	badhi	180,000	du
21	pa	2,000	ni	212,108	japipu

What are the following numbers in Aryabhata's notation?

kakiku   tajidu   nanu   bha

The ancient Indians had other syllabic counting systems, although none was as mind-bending as Aryabhata's. The *katapayadi* method, for example, was designed to enable users to memorize large numbers written in decimal notation. The digit 1 could be any of *ka, ta, pa,* or *ya,* the digit 2 could be either *kha, tha, pha,* or *ra*; and each of the other digits—3, 4, 5, 6, 7, 8, 9, and 0—also had multiple options, each a syllable containing the vowel sound *-a.* The number 1111, say, could be written using any combination of the correct syllables—i.e., *kakakaka,* or *katakata,* or, preferably, *katapaya,* since you might forget how many *ka*s there are in *kakakaka,* but you're unlikely to forget how to say *katapaya.*

In the late Middle Ages, Europe had three competing number notations: Roman numerals, Arabic numerals, and a secret system used by an order of Catholic monks.

The Cistercians were once the fastest-growing and most influential religious movement in western Europe. At its peak in the fifteenth century, the order had more than 500 monasteries from Scotland to Sicily, and from Portugal to Poland. Cistercians observed a strict ascetic lifestyle that emphasized self-sufficiency. Monks were required to work in the fields and partake in other forms of manual labor, such as building monasteries and brewing beer. As a result, the Cistercians introduced many new technologies to agriculture, engineering, and architecture.

The order was also numerically innovative. The monks had their own secret number notation, which they used across their network to denote, among other things, page numbers and dates. Cistercian notation could represent every whole number from 0 to 9,999.

26

MONKY PUZZLE

Each of the following four symbols is a number in Cistercian notation:

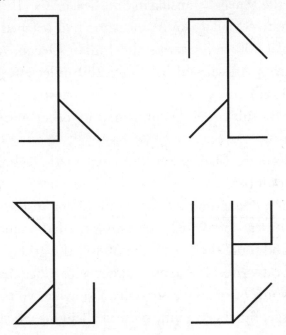

The same numbers in a different order are:

1368, 1410, 4173, and 5750

What number is this symbol?

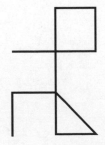

The island of Madagascar occupies a cherished place in the hearts of linguists, because it shows how studying languages can reveal unexpected historical events.

In the sixteenth century, European travelers noticed that Malagasy, the language of Madagascar, is related to the languages of the Malay Peninsula in Southeast Asia. This suggested that the native Madagascans' ancestors had reached the island by sailing 4,000 miles across the Indian Ocean, rather than arriving from Africa, which is only 250 miles away. Much to the delight of language buffs, the Asian ancestry of many Madagascans was subsequently confirmed by other means, such as DNA testing.

As a tribute to Madagascar, the next puzzle features a Malagasy crossnumber.

Clues in a crossnumber describe numbers, just as in a crossword clues describe words. For example, if the clue is "Three hundred and four" then you write 304 in the grid. If the clue is "Eleven" you write 11. A crossnumber with clues that are basic number words is, of course, trivially easy when you understand the language. But it gets a little more challenging when you don't.

27

DOWN AND ACROSS IN ANTANANARIVO

Fill in the crossnumber shown below. The clue numbers are presented in Malagasy in alphabetical order. None of the answers has zero as a first digit. If you're struggling, read the tips on the following page.

1	2		3	4
5		6		
7				
8				
9			10	

ACROSS

Dimy	*Efatra amby enimpolo sy telonjato sy sivy alina*
Fito	*Dimy ambin'ny folo sy roanjato sy arivo sy fito alina*
Folo	*Iraika amby fitopolo*
Iray	*Fito amby fitopolo*
Sivy	*Folo*
Telo	*Fito ambin'ny folo*
Valo	*Valo amby enimpolo sy sivinjato sy dimy arivo sy alina*

DOWN

Efatra	*Iraika amby valopolo sy dimanjato sy efatra arivo sy fito alina*
Enina	*Sivy amby roapolo sy telonjato*
Iray	*Iraika ambin'ny folo sy fitonjato sy sivy arivo sy fito alina*
Roa	*Dimampolo sy zato sy fito alina*
Telo	*Fito amby enimpolo sy zato sy enina arivo sy alina*

The numbers from 1 to 10 in Malagasy, in alphabetical order, are: *dimy, efatra, enina, fito, folo, iray, roa, sivy, telo,* and *valo*. That much is clear, since these are the numbers of the clues, and on the grid there are cells marked from 1 to 10. In fact, you get the value of one of these numbers for free, by realizing that one number appears on its own both in the list of clues and in the list of clue numbers. By a process of elimination, you will realize that this number can only fit in one position in the grid, which gives you a foothold to solve the rest of the puzzle. To do that, you will need to use Sudoku-like techniques, such as keeping tabs on the number of possibilities for each cell, along with arithmetical common sense and some linguistic savvy. You can assume that Malagasy numbers are decimal—so, for example, a five-digit number is described by giving the exact number of units, tens, hundreds, thousands, and ten thousands, as in English. You can also assume that shorter phrases generally describe shorter numbers. The beauty of this puzzle is in seeing how the interplay between the digits in the clues and those in the clue numbers slowly reduces your options for what digits can go where. By the end of the puzzle you will know how to describe Malagasy numbers from 1 to 99,999.

Malagasy belongs to the Austronesian family of languages, which originated in Taiwan. About 5,000 years ago, the earliest Austronesian speakers began to migrate from Taiwan. They were such skillful boatbuilders and navigators that over the following millennia they settled on islands in an area that spanned two thirds of the world's circumference, from Madagascar in the west to Easter Island in the east. About 1,200 living languages are descended from Proto-Austronesian, and their geographical span is the largest of any language family whose speakers migrated before the European Age of Exploration.

In around 800 BCE, Austronesian speakers reached the Pacific islands of Samoa. After a gap of a thousand years or so, during which they didn't sail beyond this point, the Samoans began a spectacular and speedy expansion across Polynesia, becoming the first settlers in Hawaii (2,500 miles north), Easter Island (2,000 miles east) and New Zealand (2,000 miles south).

Some Polynesian languages are notable for the small number of consonants they contain. Here's a meaningful sentence in Maori, for example, with no consonants at all:

i auee au i aa ia i aa ai i aua ao
(I wept while he drove away those clouds)

Despite such huge distances between their speakers, Polynesian languages have strikingly similar basic vocabularies. In each of the languages in the next question, for example, the words for 1 to 9 have *exactly* the same vowel sounds. The only differences are in the consonants, which change between languages in a perfectly regular way. For example, if a "k" in language A becomes a "t" in language B, you can assume that all "k"s in A become "t"s in B. (Remember Jacob Grimm: Sound shifts are systematic whether you are in Frankfurt or Fiji.)

28

PADDLING THE PACIFIC

The following table shows the numbers from one to nine in five Polynesian languages. Nuku Hiva is spoken in the Marquesas Islands and Rarotongan in the Cook Islands.
  Fill in the gaps.

	one	two	three	four	five	six	seven	eight	nine
Hawaiian	kahi	lua		ha	lima	ono	hiku	walu	
Maori	tahi	rua	toru	wha		ono	whitu	waru	iwa
Nuku Hiva	tahi		to'u	ha		ono		va'u	
Rarotongan	ta'i			'a	rima	ono	'itu	varu	iva
Samoan	tasi	lua			lima	ono	fitu		iva

Note that *wh*, and the glottal stop marked by the apostrophe ('), are single consonants.

We finish our numerical tour of the world with two word-based puzzles that reveal nonstandard ways to count.

The "base" of a number system is the main cycle of that system. The Arabic numeral system has a base of ten, since once we pass ten we start from the beginning again: 11, 12, and 13 are "ten plus one," "ten plus two," and "ten plus three," and so on. Ten is by far the most common base in the world, because humans have ten fingers. We learn to count using our fingers, so using them to set the limit of the cycle is convenient. In the highlands of Papua New Guinea, however, many communities use more than their fingers (or toes) for counting: They assign numbers to other body parts, such as eyes, elbows, and, in some cases, nipples and genitals. As a result, many of the languages spoken in Papua New Guinea have unusual bases. One is Tifal, one of the Ok languages of the Trans–New Guinea language family.

OK COMPUTER

Here are some equations in which the numbers are written in Tifal. The symbols ×, +, and = have their normal meanings—that is, "times," "plus," and "equals." None of the numbers is higher than 30.

asumano × aleeb = bokob
asumano × ataling = tadang
bokob × ataling = ataling madi
bokob × asumano = nakal madi
asumano × feet = feet madi
ataling × ataling = tadang madi
asumano + ataling = feet
feet + miit = feet madi
tadang + ataling = tadang madi

a) Write the following equations in numerals:

beeti + nakal = beeti madi
bokob + maakob = feet
awok × awok = asumano madi

b) Write the values of x and y in Tifal.

tadang + miit = x
ataling madi − aleeb = y

HINT: Try focusing on *ataling × ataling = tadang madi* to begin with.

Remember the Celtic languages and their idiosyncratic spellings (from problems 12 and 13)? Celtic numbers also have some strange quirks. In Welsh, the traditional way of saying 18 is *deunaw*, or "twice 9." No other language does *that*.

Our final number conundrum concerns Manx, a close relative of Irish and Scots Gaelic spoken on the Isle of Man. Manx was declared extinct in the 1970s when its last native speaker died. Yet it has come back from the dead, and there are now dozens, if not hundreds, of native speakers. A primary school on the island in which all lessons are taught in Manx has about 70 pupils. At this school, the children are as likely to use Manx as English in the playground.

Manx numbers are very confusing to English speakers who are used to a decimal system, although I am told that Manx children don't struggle with them at all.

30

CELTIC COUNTING

Here are nine numbers written in the Manx language, listed alphabetically:

daa-yeig

hoght-jeig as feed

jees as daeed

kiare feed as tree-jeig

nuy

queig as feed

shiaght-jeig as daeed

tree feed as kiare-jeig

tree feed as shey

And here are the numbers they correspond to, in numerical order:

9

12

25

38

42

57

66

74

93

Match the numbers written in Manx to their correct values.

This problem is difficult because, unlike the previous questions, it cannot be solved by logic alone. You will need to make intuitive guesses about how the number-words might have changed over time. An equivalent observation in English would be deducing that "thirteen" was, at one time, "three-ten." Manx is one of the Celtic languages, and you already know that they do funny things to initial consonants. Best of luck.

I started this chapter with a description of the world's oldest surviving written documents. Those clay tablets had numbers on them. They also displayed other symbols, which we'll take a look at in the next chapter.

LINGO BINGO

Chinese Compound Words

In Chinese, a compound word is a word made up of two or more characters. For example, the Chinese for "suit" is 西装, which literally means "western costume." For each of the compound words below, I've given the literal translations of the individual characters. Which of the four options is the correct meaning of the whole word?

1. BAG RAT 袋鼠

   a) Kangaroo
   b) Gopher
   c) Capybara
   d) Bat

2. ELECTRIC BRAIN 电脑

   a) Robot
   b) Anxiety
   c) Computer
   d) Power station

3. STINKY WEASEL 臭鼬

   a) Ferret
   b) Skunk
   c) Pine marten
   d) Polecat

4. COOL SHOES 凉鞋

   a) Trainers
   b) Bare feet
   c) Winklepickers
   d) Sandals

5. PEOPLE MOUNTAIN PEOPLE SEA 人山人海

   a) Archaeologist
   b) Backpacker
   c) China
   d) Crowd

6. FIRE CHICKEN 火鸡

   a) Turkey
   b) Drumstick
   c) Dragon
   d) Shirker

7. RAFTER GENTLEMAN
   梁上君子

   a) Burglar
   b) Scaffolder
   c) Gymnast
   d) Carpenter

8. RIVER HORSE 河马

   a) Buffalo
   b) Manatee
   c) Hippo
   d) Otter

9. FIRE CHARIOT 火车

   a) Motorbike
   b) Car
   c) Train
   d) Bus

10. CAT HEAD EAGLE
    猫头鹰

    a) Falcon
    b) Owl
    c) Buzzard
    d) Bat

11. OLIVE BALL 橄榄球

    a) Retirement party
    b) Rugby ball
    c) Jealousy
    d) Gnocchi

12. BEAR CAT 熊猫

    a) Panda
    b) Lion
    c) Grizzly bear
    d) Ocelot

4
Decipher
Yourself!

CRACKING
ANCIENT CODES

The deciphering of an ancient script is perhaps the most romantically alluring puzzle that exists in the real world. Fragments of unknown writing are imbued with magic and mystery. Who isn't curious to discover what they say? Once we make sense of a lost language, we can transport ourselves into the past and glimpse inside our ancestors' minds for the first time.

Successful decipherments rank among humankind's greatest intellectual triumphs. Yet not only did these achievements help unlock the mysteries of the ancient world, they also helped linguists understand the evolution of the written word. In this chapter I will chart the history of writing from the clay tablets of Sumer to the printed Latin letters you are reading on this page. Along the way you will have the chance to replicate two famous decipherments—the cracking of Old Persian cuneiform and Egyptian hieroglyphs.

The story begins in the ancient city of Uruk, on the banks of the Euphrates. There, archaeologists discovered the clay tablets displaying the pictures and numbers I mentioned in the last chapter.

31

GRIPPING REEDS

The following symbols appeared on Sumerian clay tablets dating from around 3000 BCE. Match them to their correct meanings, listed alphabetically below.

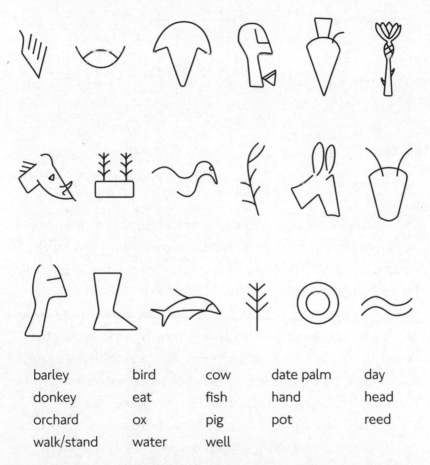

barley	bird	cow	date palm	day
donkey	eat	fish	hand	head
orchard	ox	pig	pot	reed
walk/stand	water	well		

These symbols are essentially emojis, pictographic representations of common objects. Over the next few hundred years, Sumerian scribes began to standardize and simplify their writing style. They began to approximate the picture-symbols using wedge-marks, made by jabbing the reed lightly into the clay. (Which is why the writing system is called "cuneiform," from the Latin *cuneus*, meaning "wedge.") The symbols underwent a process of abstraction, and, at some stage, many rotated a quarter turn.

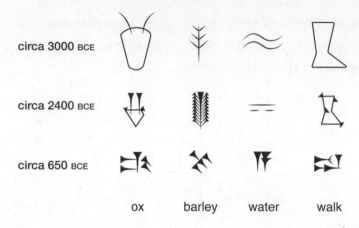

ox	barley	water	walk

The development of cuneiform. Pictographic (top); early cuneiform (middle); late Assyrian (bottom). Each wedge mark is an indentation, represented here by a long line with a triangle at the end of it. *Source: Reading the Past, 1990, British Museum Press.*

After the fall of Sumer at the end of the third millennium BCE, Assyria began its rise to become the predominant civilization in Mesopotamia. At the beginning of the first millennium BCE it was the largest empire the world had ever seen. The Assyrians spoke Assyrian and Babylonian and wrote using a cuneiform script. This version of cuneiform was mostly a syllabary—that is, a system in which each symbol represents a syllable, as opposed to an alphabet, in which each symbol represents a sound. The Assyrians used cuneiform for many purposes, not just to jot down stock inventories, but also to record recipes, hymns, rituals, prayers, and stories (such as the *Epic of Gilgamesh*, for example, the earliest major work in world literature).

In the Assyrian capital Nineveh (near present-day Mosul), the scholarly king Ashurbanipal assembled a library of cuneiform tablets, the largest collection of written material at the time. When the Assyrian empire eventually collapsed, Nineveh was burned to the ground. The blaze was clearly a tragedy for Assyrians, yet it was a stroke of luck for the library, since it flame-roasted the clay tablets, making them among the best-preserved documents of the ancient world.

The library is the single most important source for reconstructing Babylonian and Assyrian culture, and is also the source of the words in the following problem.

32

SHUT UP, SON!

Below are eight Babylonian words in cuneiform and their pronunciations (and meanings) listed in random order. Match the cuneiform with their correct pronunciations.

Cuneiform	Pronunciation and meaning
1.	a) *maru* son
2.	b) *ruqu* distance
3.	c) *qulu* silence
4.	d) *lushepisamma* I will get someone to do
5.	e) *ubla* she brought
6.	f) *lanu* form
7.	g) *nubalu* chariot
8.	h) *balu* without

In cuneiform, write *sheru,* which means "morning," and *qula,* which means "shut up!" when addressed to a man.

Cuneiform was used to write Sumerian, Assyrian, Babylonian and a dozen other languages. The most recent surviving tablet is an astronomical text from 75 CE, which means that cuneiform was in use for three millennia—a longer period of time than the Latin alphabet has been in existence.

At the end of the first century CE, however, cuneiform was obsolete, all knowledge of it forgotten. The desire to read it again only returned in the seventeenth century, when European travelers noticed the script at Middle Eastern archaeological sites. At first some scholars didn't believe it was a type of writing, suggesting instead that it was ornamental or the tracks of birds.

In the following pages I am going to take you through the first deciphering of a cuneiform script, a spectacular achievement all the more remarkable because it was made by a German schoolteacher, Georg Friedrich Grotefend. In fact, I am going to present you with the material Grotefend used to make his breakthrough, in 1802, and ask you to repeat his decipherment.

The text that Grotefend had available to him was carved into walls that were discovered in the ruins of Persepolis, the ceremonial capital of the Achaemenid Empire. The Achaemenids— the ancestors of present-day Iranians—spoke Old Persian. Their early monarchs were well known, and came from two families, whose family trees are shown opposite. Cambyses was *not* a king, but his son Cyrus and grandson Cambyses were. Likewise, Hystaspes was *not* a king, but his son Darius and grandson Xerxes were. Forgive me for this information dump, but it's all relevant to the solution of the problem.

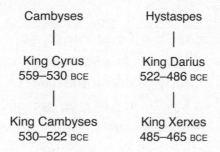

Cambyses	Hystaspes
|	|
King Cyrus	King Darius
559–530 BCE	522–486 BCE
|	|
King Cambyses	King Xerxes
530–522 BCE	485–465 BCE

The Achaemenid kings with the dates of their reigns.

The other relevant piece of information that Grotefend knew was that the leaders of the Sassanian Empire, who ruled Persia in the first century CE and saw themselves as the sucessors of the Achaemenids, often titled themselves as follows:

"X, the great king, the king of kings, the son of Y . . ."

33

THE KINGS OF OLD PERSIA

The following texts were discovered in the ruins at Persepolis, and date from somewhere between the sixth and the fourth centuries BCE. Each was carved above a figure on a doorway.

Text 1

𒀭 𒈨𒌷 𒅗 𒂍 𒅗 𒇻 𒍑 𒀸 𒁹 𒍑 𒈨 𒅗 𒆠 𒀭 𒅗 𒀸 𒂍 𒐊 𒅗
𒁀 𒀸 𒁹 𒍑 𒈨 𒅗 𒆠 𒀭 𒅗 𒀸 𒁹 𒍑 𒈨 𒅗 𒆠 𒀭 𒅗 𒈨 𒌋
𒈨 𒄭 𒀸 𒀭 𒀭 𒍑 𒁹 𒈨 𒂍 𒈨 𒆳 𒅗 𒈨 𒀸 𒈨 𒇻 𒈨

Text 2

𒁹 𒍑 𒅗 𒈨 𒂍 𒍑 𒈨 𒀸 𒁹 𒍑 𒈨 𒅗 𒆠 𒀭 𒅗 𒀸 𒂍 𒐊
𒂍 𒁀 𒀸 𒁹 𒍑 𒈨 𒅗 𒆠 𒀭 𒅗 𒀸 𒁹 𒍑 𒈨 𒅗 𒆠 𒀭 𒅗
𒈨 𒌋 𒈨 𒄭 𒀸 𒀭 𒈨 𒂍 𒅗 𒂍 𒈨 𒇻 𒍑 𒀸 𒁹 𒍑 𒈨
𒅗 𒆠 𒀭 𒅗 𒈨 𒅗 𒈨 𒀸 𒈨 𒇻 𒈨

Translate the inscriptions.

You may have just burst out laughing at the difficulty of this problem. I did, too, when I first saw it. But it can be solved using pattern-recognition skills, some logical thinking, and a few flashes of insight. It may be one of the hardest problems in this book. I decided to include it, however, because of its historical significance, and because it gives you a real sense of the extreme intellectual efforts that deciphering requires.

If you are brave, please attempt it with no help.

Or if you want your hand held, read on.

The first thing to notice is that all the symbols are made up of only horizontal and vertical wedges, apart from a single diagonal mark, ⟍, which appears after every half dozen or so signs. Grotefend guessed—correctly—that these recurring diagonals are dividers, separating the words in the text. He also guessed that the script read left to right, and from top to bottom. Again, he was right.

We want to look for patterns to see if we can simplify the text as much as possible. You may have spotted that several signs appear more than once. We'll find the text easier to read if we number each sign and rewrite it using a numeral. I've done this for you below: 1 is the first symbol that appears, 2 is the second, and so on. I have also used hyphens to link the symbols that appear in the same words.

Text 1:
1–2–3–4–5–6–7
8–7–2–4–9–10–4
5–11–3–12
8–7–2–4–9–10–4
8–7–2–4–9–10–4–2–13–2–14
15–10–7–16–2–17–18–19–4–2
18–6–20

Text 2:
8–7–4–2–3–7–2
8–7–2–4–9–10–4
5–11–3–12
8–7–2–4–9–10–4
8–7–2–4–9–10–4–2–13–2–14
1–2–3–4–5–19–6–7
8–7–2–4–9–10–4–19–4–2
18–6–20

With the text written in numbers, it's easier to spot patterns. Not only do certain symbols repeat, but several strings of symbols repeat. For example, the 1–2–3–4–5 from the first word in Text 1 reappears in the sixth word of Text 2. Let's call this string A. The second word of Text 1, 8–7–2–4–9–10–4, appears twice more in Text 1 and four times in Text 2. Let's call this string B.

I have carried on through the texts substituting strings for individual letters. If a string appears at the beginning of a word, or is a complete word, I've given it a capital letter. If a string only appears as a word ending, I've given it a lowercase letter. The list of substitutions is:

A = 1–2–3–4–5
B = 8–7–2–4–9–10–4
C = 5–11–3–12
D = 15–10–7–16–2–17–18
E = 18–6–20
F = 8–7–4–2–3–7–2
-m = –6–7
-n = –2–13–2–14
-p = –19–4–2
-q = –19–6–7

Which simplifies the inscriptions to the following:

Text 1:   A-m B C B B-n D-p E
Text 2:   F B C B B-n A-q B-p E

Each one of these texts is a statement of the form:

"X, the great king, the king of kings, the son of Y . . ."

I leave the final part of the problem to you.

Grotefend had cracked a 2,000-year-old code. Scholars built on his work, and in the following decades were able to decipher the rest of Old Persian cuneiform, which historians now believe was a script invented by King Darius purely for the purposes of making monumental inscriptions that celebrated his own magnificence. One of these inscriptions is 50 feet (15 m) high and 80 feet (25 m) long, carved 325 feet (100 m) up a cliff face in Behistun, western Iran. When this mega-inscription was transcribed—by a young British army officer perched precariously on a ladder—it was discovered to have text in both Old Persian and Babylonian cuneiform. Scholars used their knowledge of the Old Persian to decipher the Babylonian, and once Babylonian was understood, the discovery of bilingual Babylonian/Sumerian tablets in the library at Nineveh led to the decipherment of Sumerian cuneiform.

Assyriologists now consider cuneiform well understood, although due to the volume of surviving tablets and fragments—about a million in total, including 130,000 in the British Museum—many tablets are still awaiting their first reads. "In this field you still make astonishing discoveries. Things you never imagined," says Jonathan Taylor, assistant curator of the British Museum's cuneiform collection. "We find new kings. We find entire new kingdoms. Lots of new words, new types of text, new bits of literature, whole books that have been lost."

———•———

At the same time that the Mesopotamians were poking pieces of mud with sticks, one thousand miles to the west the Egyptians began to write down their own language, mostly by using ink on papyrus. The Egyptian style consisted of pictorial symbols, just like Sumerian cuneiform in its early stages. But whereas cuneiform characters lost their pictorial nature, Egyptian characters never did, and as a result, hieroglyphs are among the most bewitching of all the ancient scripts.

Knowledge of how to read hieroglyphs had been lost for well over a thousand years when, in the early nineteenth century, scholars began to try to make sense of them again. The decipherment was achieved in 1822 by the Frenchman Jean-François Champollion, who established that hieroglyphs were a mixed system in which the symbols could represent consonants, vowels, syllables, or whole words.

Champollion was able to make his breakthrough thanks to two bilingual inscriptions written in both hieroglyphs and ancient Greek: the Rosetta stone and the Philae obelisk. (The inscription on the Rosetta stone is actually trilingual, featuring a third version in another Egyptian hieratic script.)

One feature of both inscriptions was the use of an oval, or "cartouche," containing groups of symbols. Scholars had guessed correctly that the cartouches contained the names of Egyptian leaders, and that the pronunciation of those names would be more or less the same in Egyptian as in Greek.

Champollion was able to deduce a name on the Philae obelisk based on a name on the Rosetta stone. Maybe you can do the same.

34

CHAMPAGNE FOR CHAMPOLLION

The following cartouche from the Rosetta stone shows the name PTOLMES (Ptolemy).

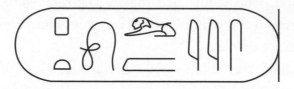

What is the name in the following cartouche from the Philae obelisk?

(The ⌒ and the ⌐ have the same meaning.)

I have adapted the Philae cartouche very slightly to eliminate extraneous detail. In fact, all printed representations of hieroglyphs involve some kind of modification to ease understanding. Egyptologists print hieroglyphs horizontally and from left to right, which is not how they generally appear in the original texts. The Rosetta stone reads right to left, for example, and the Philae obelisk reads vertically from top to bottom.

Hieroglyphs are utterly captivating. Allow me the liberty of presenting a second cartouche conundrum. You may recognize the ankh, the Egyptian symbol for life. It is a popular symbol in jewelery and tattoos that is used to represent, among other things, Africanness, Paganism, Goth subculture, and certain types of Christianity.

35

DEATH ON THE NILE

Decipher the meaning of the cartouche below, using the partial key. The cartouche was discovered on a box in a tomb in the Valley of the Kings. The figures are shown as they appear in the original, not rearranged left to right.

▭	game board with playing pieces, pronounced *mn*
∿∿∿	water, pronounced *n*
𓅨	chick, pronounced *u*
☥	cross with loop, meaning "life," pronounced *ankh*
⌐	shepherd's crook, meaning "ruler"
▯	column, meaning "Heliopolis"
⚘	heraldic plant of Upper Egypt, meaning "Upper Egypt"

Egypt was the location of a game-changing event in the history of writing: the invention of the alphabet.

An Egyptian scribe writing down a foreign name or a rare word would spell it out using hieroglyphs that represented only consonants. Around the beginning of the second millennium BCE, a group of migrants living in Egypt but speaking another language, probably a relative of Babylonian, copied this spelling-by-consonants idea but rejected the rest of the apparatus: No symbols that represented syllables. No symbols for words or phrases. The new script was an ultrasimplified, dumbed-down, bottom-of-the range imitation of Egyptian hieroglyphics, and it was also the world's first alphabet.

Writing was invented independently in Sumer, in China, in Central America, and maybe in other places too. But the alphabet was invented only once, by migrants in Egypt, and all the other alphabets of the world are descended from it.

The advantage of an alphabet, in which each symbol represents an individual unit of sound, over a syllabic system like Mesopotamian cuneiform, or a mixed system like Egyptian hieroglyphs, is that the alphabet requires vastly fewer symbols. It's easier on the memory and quicker to learn, and, as a result, the idea began to spread.

By around 1000 BCE the alphabet had arrived in the eastern Mediterranean. The Phoenicians, who lived more or less where Lebanon is now, used an alphabet of 22 consonants. The Phoenicians were an entrepreneurial, seafaring people who set up trading colonies as far away as Spain and Morocco. Their signature export was a reddish-purple dye made from the mucus of a now-extinct sea snail. (Hence their name, which comes from the Greek for "purple." The dye could be afforded only by the rich, which is why purple became associated with royalty.) But the Phoenician export that left the greatest legacy was their script, which is the origin not only of the Greek and Latin alphabets, but also of the alphabets used by Hebrew, Arabic, and hundreds of other languages spoken across Asia as far as Mongolia and Bali.

When you tackle the next problem, look for similarities between Phoenician letters and Latin ones, and remember that the Phoenician alphabet contains only consonants.

36

PURPLE REIGN

The eight ancient cities listed below in the Phoenician alphabet are also marked on the map with their English spellings, which reflect the Phoenician pronunciation. The darker-grey area shows the boundaries of the Phoenician civilization around 1000 BCE.

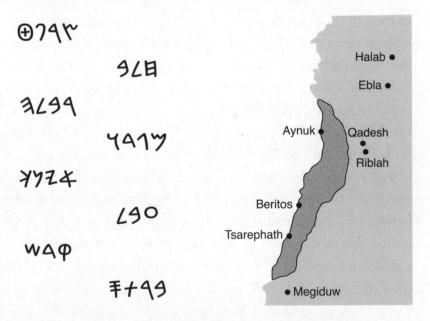

Match the cities to their Phoenician names. Which of these cities is still a regional capital today?

Before we move on to the ancient Greek alphabet, a detour to Crete. In a chapter about decoding ancient scripts, it would be remiss of me not to mention Linear B, the most famous decipherment of the twentieth century.

For 50 years the greatest mysteries in epigraphy were two types of writing unearthed on Crete, among the ruins of Knossos, in 1900. The text appeared on clay tablets dating from the second millennium BCE, making them Europe's oldest written documents. Both scripts, named Linear A and Linear B, were made up of intricate lines. Linear B seemed to be a refined version of Linear A: Its symbols were intricate and beautiful, like nothing ever seen before. Some were clearly pictograms, such as ⵛ (deer), ⵛ (horse), and ⵛ (wheeled chariot), but the script also included about 80 nonpictographic signs, including Φ, ⊕, and ⵛ. The decipherment of Linear B captured the imagination of the public and scholars alike, yet it posed a seemingly impossible challenge. Not only did no one know what the signs stood for, no one knew what language the symbols were written in.

Alice Kober, a classics professor in New York, prepared her own library of 180,000 index cards in order to analyze statistically the combinations in which the signs appeared. Building on her work, in 1952 the British architect Michael Ventris completed the decipherment. He was 29 years old and had been obsessed with Linear B since seeing the script in a museum when he was 14. Ventris confirmed that Linear B was a syllabic script, written left to right, which was used to write Mycenaean Greek, a dialect of Ancient Greek that was spoken around 1500 BCE, thus predating, by 1,000 years, what we commonly refer to as Ancient Greek, the Greek of Athens spoken around 500 BCE. The decipherment was a dazzling intellectual feat, and it has enabled historians to open up a forgotten world.

37

A CRETAN CRUNCHER

Below is a list of words in Ancient Greek. The symbols in the first column show how they would have been written using Linear B. The second and third columns contain the same Ancient Greek words (written using the Latin alphabet) and their English translations, in random order.

Match the Linear B to the Ancient Greek.

Linear B		Ancient Greek		meaning
1.	⊤ᛞ†	a)	Knossos	Knossos
2.	ᛁᛁᛁᛁ	b)	dōra	gifts
3.	ᛗᛗ	c)	heneka	because of
4.	⊕ᛞ	d)	epi	on top of
5.	‡᛬	e)	thygatēr	daughter
6.	ᛁᛁ	f)	meli	honey
7.	‡ᛗᛞ	g)	meta	in order to
8.	ᛁᛁᛁᛁᛁ	h)	para	for (someone)
9.	⌀⊕᛬	i)	patēr	father
10.	ᛗᛞ⊕	j)	tripodes	three-legged
11.	ᛁᛞᛁ	k)	pharmakon	drug
12.	ᛁᛁ	l)	chalkon	copper
13.	‡ᛁᛁ	m)	chrysos	gold
14.	ᛞᛁᛁᛁ	n)	agros	field

Feeling stumped? You may well be. Linear B is a syllabary, meaning that every symbol represents a syllable. Yet the puzzle contains more Ancient Greek syllables (26) than it does Linear B symbols (23). Some symbols, therefore, must refer to more than one syllable. Clue: one symbol is pronounced *pa*, *pha*, and *ba*.

The reason some symbols have multiple pronunciations is that Linear B was used to write Mycenaean Greek, a language spoken many centuries before Ancient Greek, and the symbols of Linear B cannot distinguish between all the sounds that the Ancient Greek (and Latin) alphabet can.

Usually, if a symbol can be pronounced in different ways, only the consonant will vary in the Ancient Greek transliterations, and these consonants will be phonetically similar. (Two consonants will sound similar if your lips and tongue are in similar positions when you pronounce them.)

Back to Ancient Greek, and the puzzle version of a palate cleanser. For the first time in this chapter, no funny-looking characters involved!

38

MASTERS AND SLAVES

Match the phrases in Ancient Greek opposite (transcribed using the Latin alphabet) to their correct translations.

(The letter ō is a long o.)

1. the donkey of the master

a) *ho tōn hyiōn dulos*

2. the brothers of the merchant

b) *hoi tōn dulōn cyrioi*

3. the merchants of the donkeys

c) *hoi tu emporu adelphoi*

4. the sons of the masters

d) *hoi tōn onōn emporoi*

5. the slave of the sons

e) *ho tu cyriu onos*

6. the masters of the slaves

f) *ho tu oicu cyrios*

7. the house of the brothers

g) *ho tōn adelphōn oicos*

8. the master of the house

h) *hoi tōn cyriōn hyioi*

The Ancient Greek alphabet was different from the Phoenician in one important way—it included vowels as well as consonants:

Α Β Γ Δ Ε Ζ Η Θ Ι Κ Λ Μ Ν Ξ Ο Π Ρ Σ Τ Υ Φ Χ Ψ Ω

In this way, the Ancient Greek alphabet is the first "true" alphabet, in that vowels and consonants are treated equally.

The Greek alphabet spread to the Italian peninsula. It was adapted by the speakers of the many languages that were spoken there, such as Etruscan, Umbrian, and Oscan, and ultimately Latin, which became the standard as its speakers, the Romans, conquered all their neighbors, and then their neighbors' neighbors, and beyond. The Latin alphabet disseminated across the Roman Empire, becoming over time the most-used alphabet in Europe, and then the world.

When we think about how the Latin alphabet evolved from the Greek one, the Etruscan, Umbrian, and Oscan alphabets help us to fill in some of the gaps. The next problem involves Oscan, which used to be one of the most widespread languages in central and southern Italy, and which was still in active use around 100 BCE, around the time Cicero and Julius Caesar were born. One of the most important surviving Oscan texts is the Cippus Abellanus, a tablet from the second century BCE that documents a legal dispute between the cities of Abella and Nola, both located near Naples.

As a code-breaking puzzle, the following problem is the most Sherlock Holmesian of the book. At first, it is impenetrable. You clearly have far too little information to make any headway. You may achieve small advances, such as by presuming (correctly) that the dots at mid-height divide the text into words, and by noticing that some of the letters look familiar, such as the T, V, and backward E. Then, you will sit back in your leather chair, puff on your pipe, and, given enough time to reflect, see in a flash what you did not see before: The solution is hiding in plain sight. Full disclosure: I was unable to solve this problem when I came across it, but I gasped with delight when I read the answer. The puzzle unravels with one simple insight, which is the kind of observation a fictional detective might make when asked to decipher a piece of code.

39

AND THE OSCAR FOR OSCAN GOES TO . . .

Shown below is a section of the Cippus Abellanus, preceded by its English translation.

> Behind the walls which go around the sanctuary, in this area neither the inhabitants of Abella nor the inhabitants of Nola [are permitted to build] anything.

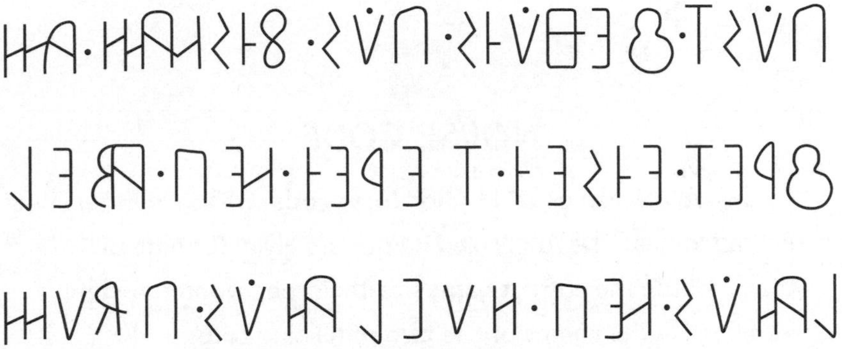

Below are 16 words that appear in the Cippus Abellanus, transcribed into the Latin alphabet. Some of them appear in the portion of inscription shown above, and some do not:

eisei	púst	prúftú	fisnam
fufans	anter	amfret	pús
svai	feihúis	pússtis	inim
ehtrad	pidum	terei	eisúd

1. Which of these words appear in the Cippus Abellanus?

2. Using the Latin alphabet, give the Oscan word for:

   a) neither/nor
   b) inhabitants of Abella
   c) inhabitants of Nola

The runic alphabets of northern Europe are derived from the pre-Latin alphabets used in northern Italy. Runes are made up mostly of vertical and diagonal strokes. This disctinctive style presumably came about because they were written on wood, and horizontal strokes would have gone against the grain. The earliest runic alphabet, Elder Futhark, was used between the second and the eighth centuries CE, mostly in Scandinavia. Its name comes from its first six letters: *f*, *u*, *th*, *a*, *r*, and *k*. (The *th* is a single letter, from which the Old English letter *ð*, or "eth," is derived.)

40

NORSE CODE

Below are the names of 11 Old Norse gods written in Elder Futhark runes. The Anglicized names are given for nine of them. Match the correct names to their runes, and translate into English the names of the two remaining gods.

1. ᛒᚨᛚᛞᚱ   a) Baldur

2. ᚦᚩᚱ   b) Dallinger

3. ᛁᚦᚢᚾᚾ   c) Day

4. ᛗᚨᚷᚱ   d) Earth

5. ᚾᚩᛏᛏ   e) Freya

6. ᚹᚱᛖᛁᚨᚠ   f) Freyr

7. ᛋᛟᚱᚦ   g) Ithun

8. ᛉᛖᛚᛁᛟᛗᚱ   h) Night

9. ᚠᚱᛖᛁᚱ   i) Sun

10. ᛟᛉᛁᛏ   j)

11. ᚲᛟᛚ   k)

Sometime in the late 1990s, a consortium of tech companies, at a meeting in the offices of the Swedish firm Ericsson, decided to name a new industry-standard wireless communication system after a tenth-century Danish king. Harald Bluetooth had united the Danes and conquered Norway. His success at connecting the Norse peoples was seen as an appropriate metaphor for connecting electronic devices.

The symbol they adopted for Bluetooth technology was ᛒ. It is a "bindrune" (a superimposition of one rune on another) of ᚼ and ᛒ, the dentally discolored monarch's initials in Younger Futhark, the descendent of Elder Futhark that was in use during his reign.

The ancient Norwegian alphabet may have been in disuse for half a millennium, yet every day, when we glance at our phone or computer screens, we are reading the runes.

LINGO (B) (I) (N) (G) (O)

Portmanteau Words

A "portmanteau" is a word made up of two separate words. For example, "muppet" is a portmanteau of "marionette" and "puppet." Derive the meaning of each of the portmanteaus below, by breaking them down into their two constituent parts.

1. BROZILIAN
   a) Beauty treatment
   b) Easy-going male
   c) Bros superfan
   d) Brazilian friend

2. CANKLE
   a) Metallic ankle
   b) Waxy ankle
   c) Swollen ankle
   d) Feline ankle

3. SPORK
   a) Bottle stopper
   b) Bird
   c) Salami
   d) Piece of cutlery

4. FLUMPET
   a) Woman of ill repute
   b) Fat person
   c) Musical instrument
   d) Character in Winnie-the-Pooh

5. INTERROBANG
   a) Punctuation mark
   b) Funeral
   c) Interview technique
   d) Global NGO

6. THROUPLE
   a) Coin
   b) Bottle opener
   c) Fruit
   d) Relationship

7. SNICE
   a) Artificial snow
   b) Compliment
   c) Small insects
   d) Grain

8. CRONUT
   a) Old woman
   b) Piece of hardware
   c) Pastry
   d) Avian genitalia

9. JORTS
   a) Japanese medicinal plant
   b) Item of clothing
   c) Keep-fit activity
   d) Skin ailment

10. MIZZLE
   a) Smoke
   b) Woodworking tool
   c) Rain
   d) Martini cocktail

5
Relative Values

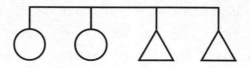

THE LANGUAGE
OF FAMILY

You can tell a lot about a society from its vocabulary of family words. English, for example, has precise words for mother, father, son, daughter, brother, and sister. But beyond the nuclear family it all gets a bit vague: A "cousin" can be any child of any aunt or uncle; a "brother-in-law" can be your wife's brother or your sister's husband. English words for family relations reflect the cultural value placed on the nuclear family in English-speaking countries, where the basic unit of society is Mom, Dad, and their 2.2 kids.

Compare this to Bengali, a language—like many in the Indian subcontinent—with a particularly lavish vocabulary of kinship terms. It has distinct words for dozens of family members, such as "younger sister's husband" and "father's elder brother's wife." In Bangladesh and India, where extended families often live together, these words are helpful in describing who's who and in identifying everyone's role. The words also clarify who's related, and therefore not a potential husband or wife.

This chapter is about the words we use for our nearest and dearest. We will explore the language of family, and a family of languages. We will visit the parched Australian outback and the coldest country in the world.

But first, a question about grammar.

Sorry, I mean grandma.

41

THE FARFAR NORTH

The following four words are common to Danish, Norwegian, and Swedish:

mormor   *morfar*   *farmor*   *farfar*

The first word can be translated as "grandmother." But usually it means something more precise.

What are the precise meanings of all of these words?

Southern Europeans also have words for family members that have no English equivalents. Going back a couple of thousand years, Latin is very precise when it comes to describing certain generations.

In the next question you have to assume you are very old (or long-deceased), although your age (or mortality) is not relevant to the answers.

(To solve the problems in this chapter, you'll need to be able to read a family tree. Double horizontal lines indicate marriage; siblings are joined by horizontal lines; vertical lines join parents, above, to children, below.)

42

MY ROMAN FAMILY

The diagram shown below is your family tree. Males are underlined. The five following statements are true. The Latin phrases in italics describe family relations.

Titus is your *avunculi abnepos*
Septimus is your *amitae pronepos*
Livia is your *materterae neptis*
Florentina is your *patrui adneptis*
Flavia is Augusta's *trineptis*

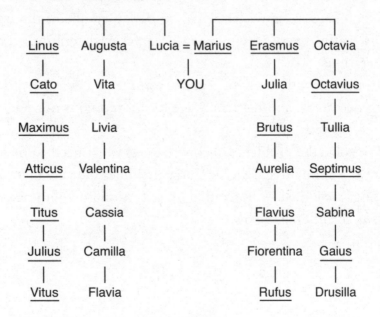

In Latin, what is the relationship of:

Camilla to Cato
Rufus to Octavius
Your daughter to Vita
Your great-grandson to Julia

Danish, Norwegian, Swedish, Latin, and Bengali—the languages mentioned so far in this chapter—are all members of the Indo-European language family. The first three are like siblings, closely related and descended from Old Norse, a language spoken in Scandinavia about a thousand years ago. Latin is a cousin a few generations removed, and Bengali an even more distant relative. Old Norse, Latin, and Bengali were descended from languages that were descended from languages that were ultimately descended from a single language spoken about six thousand years ago, probably around the northern shores of the Black Sea. This language, "Proto-Indo-European" (PIE), is also the ancestor of all Celtic, Germanic, Romance, and Slavic languages, as well as Asian languages, including Persian, Pashto, Kurdish, Urdu, and Hindi. About half of the world's population have an Indo-European language as their mother tongue.

Languages that share a common ancestor often share similar grammars and similar vocabularies. Commonly used words, such as terms for numbers, and those for close family members, are often the most resistant to change when languages diverge over time. In the following puzzle you'll see resemblances in basic vocab between Indo-European languages spoken hundreds—even thousands—of miles apart.

43

MEET THE RELATIVES

Here are ten common words in seven Indo-European languages: Breton, English, Faroese, Friulian, Hindi, Limburgish, and Upper Sorbian.

Which column corresponds to which language?

English	a	b	c	d	e	f
father	pari	faðir	tad	pitaa	nan	fatter
mother	mari	móðir	mamm	maataa	mać	moder
brother	fradi	bróðir	breur	bhaai	bratr	broor
sister	sûr	systir	c'hoar	bahan	sotra	zöster
I	jo	eg	me	main	ja	ich
you	tu	tú	te	tum	ty	doe
one	un	ein	unan	ek	jedyn	ein
two	doi	tveir	daou	do	dwaj	twie
arm	brač	armur	brec'h	baahu	bróń	erm
bird	ucel	fuglur	evn	pakshi	ptačk	voegel

Breton (which has about 200,000 speakers) is a Celtic language spoken in the westernmost tip of France's Brittany peninsula.

Faroese (70,000 speakers) is the language of the Faroe Islands, which are located between Norway and Iceland.

Friulian (600,000 speakers) is spoken in the far northeast corner of Italy near the border with Slovenia. It is taught in school, and road signs are bilingual, written in both Friulian and Italian.

Hindi (640 million speakers) is the world's third-most-spoken language, if you include people who speak it as a second language. It is one of the two official languages of the Indian government, alongside English.

Limburgish (1.5 million speakers) is a Germanic language. Most of its speakers live in the southeast Netherlands, some in the Liège area of eastern Belgium, and a few in the Rhineland area of southwest Germany.

Upper Sorbian (20,000 speakers) is a Slavic language spoken in eastern Germany. Within the Sorbian-speaking area, the language—and its sister language, Lower Sorbian—are given equal status with German.

In the previous chapter I mentioned the Austronesian language family, whose members have a common ancestor spoken thousands of years ago in Taiwan. This family includes Malagasy, which is spoken in Madagascar. And it is to this tropical island we now return.

A traditional ritual in Madagascar is the *Famadihana*, a "bone turning" ceremony in which families exhume the bones of their ancestors from the family crypt, rewrap them in fresh cloth, and then pass them around the living descendants. This ceremony may involve conversations about great-grandchildren, great-great-grandchildren, great-great-great-grandchildren, and great-great-great-great-grandchildren. Thankfully, Malagasy contains family terms that describe each of these relations in a clear and nonrepetitive way.

RICE WITH THE GRANDKIDS

Here's a list of words in Malagasy, along with their English translations. Both sets of words are listed alphabetically.

hafaladia	ankle
kitrokely	grandchild
lohalika	great-great-grandchild
mahambozona	great-great-great-grandchild
mahandohalika	great-great-great-great-grandchild
tanim-bary	knee
zafim-bary	one who can carry something on his neck
zafin-dohalika	one who can get on his knees
zafim-paladia	rice field
zafin-kitrokely	shoot of rice (departing from the stem)
zafy	up to the sole

Match the Malagasy words with their correct English translations.

To solve this problem, you will need to compare the structure (morphology) of the Malagasy words to the meaning (semantics) of the English terms. Logical deduction takes you only part of the way, though: Common sense and lateral thinking are also required. You've probably never tried to solve a problem like this before, so don't worry if you find yourself hitting your head against the wall. (I struggled too, but the solution, when you get there, is lovely.)

I'll start you off. In the left-hand column five words appear to be morphologically related: the ones beginning with *zafy/zafin/zafim*. Of these, *zafy* appears to be the root word, and all the others seem to be derived from it. Likewise, in the right-hand column, four of the phrases are semantically related: the ones containing the word "grandchild." Of these, "grandchild" is the root word. So, tentatively, we can match our first pair: *zafy* = "grandchild."

We're left with three generations of grandchild, but four *zafin/zafim* words. In other words, three of the following terms describe a particular generation of grandchildren, but one of them doesn't:

zafim-bary
zafin-dohalika
zafim-paladia
zafin-kitrokely

We're looking to find the word that does not mean a particular generation of grandchildren, and yet contains the root "grandchild." What kind of word might this be? Well, it probably has a meaning that relates to being a descendant of someone or something. Do any of the English phrases fit the bill?

Now, on to a well-known English idiom that describes a male ancestor of the previous generation. An affectionate phrase for "father" is "the old man." As in:

"The old man gave me a chair for my birthday."

Here's a grammatical sentence that starts with the same three words:

"The old man the boat."

Huh? Linguists call this type of construction a "garden-path" sentence, because the syntactic ambiguity tricks you. The sentence leads us down the garden path: On first reading, we assume the subject is an old man, or a father, until we get to the end and realize we've been duped. The word "man" is not a noun, as we had assumed, but a verb. The old people are manning the boat.

Garden-path sentences are often a source of humor. Time flies! (You can't. They go too fast.) The following puzzle invites you to create your own.

45

THE COUSIN WHO HUNTS DUCKS

Using each of the following phrases, create two sentences with different sentence structures: an obvious one, and a less obvious one. For example:

The old train . . . The old train is broken

The old train the young

The structure of the first sentence is [The old train] [is broken], and the structure of the second is [The old] [train] [the young]. In the first one "train" is a noun, and in the second it is a verb. Off you go:

1. The cousin who hunts ducks . . .
2. The florist sent the flowers . . .
3. The cotton clothing . . .
4. The woman who whistles tunes . . .
5. We painted the wall with . . .
6. I convinced her children . . .
7. When the baby eats food . . .
8. Mary gave the child the dog . . .
9. The girl told the story . . .
10. That John is never here . . .

Let's stop walking down garden paths and return to the theme of family. The indigenous people of Australia number about 800,000 and they speak about 160 languages. (This linguistic diversity is decreasing rapidly, however, and only the 13 languages that are currently spoken by children have a chance of surviving to the end of this century.) Aboriginal communities traditionally have no kings, no elected rulers, nor any formal apparatus of governance. Instead, their rules of social organization are based around family relations. These practices are are so complex that they are studied not only by linguists and anthropologists but also by mathematicians.

One concept common to all aboriginal groups is the "classificatory" sibling. In English, my siblings are, exclusively, the other children of my parents. For an aboriginal Australian, however, "siblings" are a whole class of relation that includes:

- biological siblings
- children of maternal aunts (but not maternal uncles)
- children of paternal uncles (but not paternal aunts)
- grandchildren of maternal grandmothers (but not paternal grandmothers)
- grandchildren of paternal grandfathers (but not maternal grandfathers)

In other words, a "classificatory" sibling is anyone who is a biological sibling, or a cousin descended from a matrilineal ancestor on the mother's side, or a patrilineal ancestor on the father's side. It is therefore possible for an aboriginal man to be sitting with, say, ten brothers, none of which share any of the same parents.

The rules of behavior in aboriginal society are based around classificatory siblings. In general, an adult male is not allowed to sit with, walk alongside, or eat from the same food as any woman he calls "sister." He is also forbidden from mentioning a sister by name in any circumstance. He could refer to "the one with the long hair," but never, for example, "Amanda." Likewise, an adult female is prohibited from carrying out these activities with any of her brothers.

These strict rules about siblings are linked to taboos about marriage, and they are so fundamental they are even embedded in the grammar of some languages. The next problem concerns Murrinhpatha, which is spoken around the town of Wadeye in the Northern Territory.

46

AMY, SUE, AND BOB, TOO

Sue, Amy, Des, and Bob are siblings, and Daisy, Jane, Fred, Jim, and Dave are also siblings. The two families are not related.

In each of the following ten sentences, the people involved are "teaching each other." So, the first sentence states that Des and Bob are teaching each other.

> *Des i Bob puddeniyithnu*
>
> *Daisy i Jane puddeniyithnu*
>
> *Amy i Des puddeniyithnu*
>
> *Des i Fred puddeniyithnunintha*
>
> *Amy i Daisy puddeniyithnungintha*
>
> *Amy i Fred puddeniyithnungintha*
>
> *Des, Bob, i Fred puddeniyithnuneme*
>
> *Amy, Daisy, i Jane puddeniyithnungime*
>
> *Amy, Des, i Daisy puddeniyithnungime*
>
> *Fred, Jim, i Dave puddiniyithnu*

Complete the following Murrinhpatha sentences, so that each sentence means that the people involved are teaching each other.

1. *Bob i Jim* _____
2. *Des i Daisy* _____
3. *Bob, Jim, i Dave* _____
4. *Fred i Dave* _____
5. *Jane i Dave* _____
6. *Des, Jane, i Fred* _____
7. *Daisy, Jane, i Sue* _____

Not only are there "classificatory siblings," there are also "classificatory parents." For example, aboriginal people will use the term "mother" not only for the person who gave birth to them but also for that person's sisters.

Perhaps the most widespread social distancing rule across aboriginal Australia, however, concerns men and their "classificatory mothers-in-law"—that is, a man's mother-in-law and her sisters. Men should not be around their mothers-in-law. If, however, a situation arises in which a man and his mother-in-law must talk to each other, many aboriginal languages require both parties to talk in a particular way, such as by using special words. Sometimes this "avoidance vocabulary" contains so many special words that it is essentially a separate language. The linguist R. M. W. Dixon writes that a man and his mother-in-law communicating using the avoidance vocabulary of the aboriginal language Dyirbal must not look each other in the eye, nor stand face-to-face, and must address remarks through a third person. If no one else is present, he writes, "they might pretend to be talking to a dog or even a stone."

The system of classificatory siblings complements another widespread practice in aboriginal society, the "skin system," in which a community is divided into subsections, called skins, the purpose of which is to determine social roles, and especially whether people can or cannot marry each other. The most common system is one with four skins. Every person in that community is born into a particular skin, which is determined by the skins of their parents. In simple terms, if your mom is skin A, and your dad is skin B, then you are skin C, and you must marry someone who is a D.

Some groups, such as the Warlpiri, have eight skins. Their rules for determining who belongs to which skin are set out in the next puzzle. (Mathematicians may recognize the rule as the dihedral group of order 8, which is the group of symmetries of a square. Going from husband to wife or vice versa is like flipping the square over, and going through the four skins, 1-4-2-3-1, or 5-7-6-8-5, is like rotating the square 90 degrees each time. But I digress.)

The Warlpiri, who live about 200 miles north of Alice Springs, refer to each other most often using their skin names because the skin system works as a shorthand for the rules about classificatory siblings. For example, if a woman is skin 1, all of her classificatory brothers and sisters are also skin 1. When you know a person's skin, you know exactly the type of family relationship you have with that person, and this is important information for how you should behave.

47

MY WIFE'S FATHER'S MOTHER'S BROTHER

The Warlpiri skin system is shown in the diagram below. Each number corresponds to one of the eight skins. Horizontal rows indicate marriage correspondences, and the arrows point from mother to child.

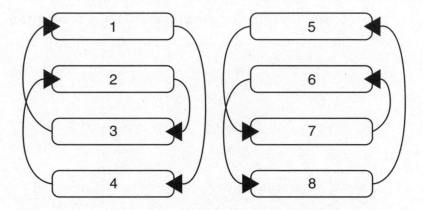

For example, if your skin is 1, you must marry someone of skin 5 (of the opposite gender) because these numbers are on the same horizontal line. If you are a female of skin 1, your child is of skin 4 since the arrow points from 1 to 4.

Males of skin 1 are called Jakamarra, and females of skin 1 are called Nakamarra. The names of the other skins always begin with a J for men, which is replaced with an N for women. (Except Jupurrula, which becomes Napurrula.)

Imagine you go into a Warlpiri village and speak to six people about their family relations. Based on the responses shown below, can you determine the name of each skin?

i) "I am a Jangala. My daughter is Nampijinpa."

ii) "I am a Nakamarra. My brother's son is Jupurrula."

iii) "I am a Nampijinpa. My mother's grandfathers were Jungarrayi and Jupurrula."

iv) "I am a Napangardi. My husband's sister's husband's father's father's mother was Napurrula."

v) "I am a Napanangka. Some of my good friends are Napaljarri and Nangala and Nungarrayi. Oh, you wanted me to talk about my family? Oops."

vi) "I am a Japanangka. My wife's father's mother's brother's wife's father's mother's brother's wife's father's mother's brother's wife's father's mother's brother's wife's father's mother's brother's wife's father's mother's brother's wife was Napurrula. I know my family tree very well."

———— ·•· ————

In Burma, many Buddhists name their children according to a rule. The next question asks you to deduce what it is.

48

BURMESE BABIES

In February 2020, the following 22 babies were born in Yangon:

Date	Day	Name	Gender
1	Sat	*thɛʔ auŋ*	M
5	Wed	*lwiŋ koko*	M
6	Thurs	*mimi khaiŋ*	F
10	Mon	*kethi thuŋ*	F
11	Tues	*zeiya cɔ*	M
11	Tues	*su myaʔ so*	F
12	Wed	*wiŋ l muŋ*	F
13	Thurs	*phouŋ naiŋ thuŋ*	M
15	Sat	*thouŋ uŋ*	F
17	Mon	*khaiŋ miŋ thuŋ*	M
18	Tues	*susu wiŋ*	F
19	Wed	*wiŋ cɔ auŋ*	M
19	Wed	*yiŋyiŋ myiŋ*	F
20	Thurs	*pyesouŋ auŋ*	M
22	Sat	*tiŋ mauŋ laʔ*	M
24	Mon	*kauŋ myaʔ*	M
25	Tues	*shaŋ thuŋ*	M
25	Tues	*shu maŋ cɔ*	F
26	Wed	*yadana u*	F
27	Thurs	*myo khiŋ wiŋ*	M
27	Thurs	*paŋ we*	F
29	Sat	*tiŋ za cɔ*	F

On the following six dates—February 3, 4, 15, 16, 26, and 27—these other children were born, but not necessarily in the order in which they are listed:

Girls

khiŋ le nwɛ

daliya

e tin

phyuphyu wiŋ

Boys

so mo cɔ

ye auŋ naiŋ

Who was born when?

(Note on pronunciation: ɛ is like "a" in "hat"; ŋ is like "ng" in "hang"; ɔ is like "a" in "hall"; and ʔ is a glottal stop, like the sound in the middle of "uh oh.")

Icelanders face a different challenge when naming their children. Parents must choose from a list of about 4,000 legally approved first names. They go from Aage to Öxar for boys, and from Aagot to Ösp for girls. (The letter *ö* is the last letter of the Icelandic alphabet.)

If you would prefer to give your child a name that's not on the list, you are allowed to submit it for approval. The Naming Committee makes its decision based on the name's compatibility with Icelandic tradition and orthography, and how embarrassing it might be for the child. In 2020, Krákur (boy) and Stormey (girl) were accepted. Lúcifer (boy) and Rosemarie (girl) were rejected, the former for obvious reasons and the latter because the correct way to write it would be Rósmary or Rósmarý.

First names are taken seriously in Iceland because Icelandic family names do not exist. One's name consists of a first name and either a matronymic or patronymic with the suffix *-dóttir* (meaning "daughter of"), either a matronymic or patronymic with the suffix *-son* ("son of"), or both. For example, if Aage and Ösp had a girl called Stormey, they might call her Stormey Aagedóttir or Stormey Aspardóttir ("Aspar" is the genitive of Ösp). If they had a boy called Krákur, he might be Aageson or Asparson. The parents are free to choose which alternative they prefer.

In centuries gone by it was also common to add an "avonymic" with the suffix *-sonar* or *-dóttur*, which are the genitive forms of *-son* and *-dóttir*. For example, Krákur Aageson Öxarssonar means Krákur, son of Aage, who is the son of Öxar. Avonymics proliferate in the old Icelandic sagas, and are now used informally.

The absence of Icelandic family surnames means that in Iceland, all lists of personal names—the telephone directory, author references in bibliographies, the national registry—are always ordered alphabetically by first name.

The only time a surname appears in an Icelandic name is if you or one of your antecedents has married a non-Icelandic person. In that case, you have the option to give your child the non-Icelandic family surname (or take it on as your married name). Even so, there are still restrictions. If someone wants to use a non-Icelandic surname as a last name (rather than as a middle name), then it must be from a list of names drawn up in 1925.

In 2019, in a quintessential example of Scandinavian progressivism, the Icelandic government announced the introduction of a new naming rule. If you are registered as gender nonbinary, you are permitted to change the patronymic/matronymic suffix to -*bur*, meaning "child of."

The following problem involves names that contain a patronymic and a matronymic together, as well as names with an avonymic.

49

A TREE IN ICELAND

Guðrún Eriksdóttir Hrafnhildardóttir and Jakob Kristjánsson had three children, and a certain number of grandchildren and great-grandchildren. To commemorate their 70th wedding anniversary, they invited all their descendants and descendants' spouses to a party. Almost everyone showed up. The party consisted of:

Daníel Guðrúnarson
Daníel Steinunnarson Guðrúnardóttur
Gunnar Gunnarsson
Ingimundur Sigurðarson Bergmann
Jakob Þorarinsson
Jón Oddsson Bergmann
Margrét Steinunnardóttir
Ragnheiður Jakobsdóttir
Rakel Ragnheiðardóttir Bergmann
Róbert Bergmann Gunnarsson
Sara Jakobsson Þorarinssonar
Sigurður Jónsson Bergmann
Stefán Gunnarsbur Gunnarssonar
Steinunn Jakobsdóttir

Assume all marriages are between a man and a woman, and that no one in this family is divorced. Margrét is 11 years old.
Draw the family tree.

As I mentioned at the start of this chapter, English stands out in global terms for its relative paucity of kinship terms. Yet you can still devise a good puzzle with a no-frills vocabulary.

50

THE FAMILY TABLE

The table shown below describes the relationships among five people in the same family.

	Alexa	Edward	Iggy	Ollie	Uzzie
Alexa	*	1	1	1	2
Edward	3	*	4	4	5
Iggy	3	4	*	4	6
Ollie	7	8	8	*	9
Uzzie	10	3	11	11	*

Each number from 1 to 11 represents a unique word from the following list:

> mother, father, son, daughter, sister, brother, uncle, aunt, nephew, niece, cousin, grandfather, grandmother, grandchild

The table is read: [*name from left column*] is the [*number*] of [*name from top row*]. Thus:

> Alexa is the 1 of Edward
> Edward is the 3 of Alexa, etc.

Alexa is female and Edward is male. Match the numbers to the correct words.

To approach this puzzle, look at the symmetries between certain relationships: For example, if X is the cousin of Y, then Y is the cousin of X. Likewise, if X is a sibling of Y, then Y is a sibling of

X. The table tells us which two family members have a symmetrical relationship, and this observation gives the way in.

———•—·—•———

Talking of families, when did you last call your mother?

Tell her you've got an appointment with the inventor of the telephone.

LINGO ⒷⒾⓃⒼⓄ

Animal Sounds

Each of the examples given below is the same animal sound written in four different languages. (In languages that don't use the Latin alphabet, the given word is a transliteration.)

Which animal do the sounds refer to?

1. *Summ*   German
   Boon   Japanese
   Zoum   Greek
   Vizzz   Turkish

   a) Bird
   b) Bee
   c) Mosquito
   d) Cheetah

2. *Pip-pip*   Danish
   Tziff-tziff   Hebrew
   Csip-csirip   Hungarian
   Fiyt-fiyt   Russian

   a) Small bird
   b) Grasshopper
   c) Mosquito
   d) Mouse

3. *Cot cot codet*   French
   Ock-ock   Swedish
   Gut gut gdak   Turkish
   Tok tok   Dutch

   a) Woodpecker
   b) Goose
   c) Fox
   d) Hen

4. *Oh ioh*   Italian
   Ia-ia   Russian
   Yi-ah   Hebrew
   A-iii a-iii   Turkish

   a) Medium bird
   b) Hyena
   c) Donkey
   d) Wolf

5. *Hov hov*   Turkish
   Guf guf   Spanish
   Gav gav   Greek
   Vov vov   Danish

   a) Dog
   b) Horse
   c) Elephant
   d) Seagull

6. *Guru guru*   German
   Burukk   Hungarian
   Guli-guli   Russian
   Gu gu gu guuk Turkish

   a) Toad
   b) Pigeon
   c) Turkey
   d) Duck

7. *Hoo hoo*   Urdu
   Hoh hoh   Japanese
   Uh! Uh! Uh! Russian
   Oe hoe   Dutch

   a) Cuckoo
   b) Nightingale
   c) Monkey
   d) Owl

8. *Pi-pi-pi*   Russian
   Squit   Italian
   Chu-chu   Japanese
   Cin-cin   Hungarian

   a) Small bird
   b) Wasp
   c) Mouse
   d) Pig

9. *Paka paka*   Japanese
   Ta-tá ta-tá ta-tá Russian
   Deg-a-dek   Turkish
   Klip klap   German

   a) Horse
   b) Woodpecker
   c) Cockerel
   d) Cicada

10. *Ko ack ack ack* Swedish
    Kuty kurutty   Hungarian
    Vrak vrak   Turkish
    Kvaak   Finnish

    a) Duck
    b) Chicken
    c) Frog
    d) Penguin

6

Aiding and Alphabetting

WORDS THAT GIVE
A HELPING HAND

The alphabet trumps all other writing systems for its efficiency. Small number of symbols; maximal number of sounds. Yet it is not a perfect, universal tool. This chapter concerns alphabets that have been redesigned, or repurposed, in the service of people with particular needs. Deaf people. Blind people. Monks. Memory champions. Typists. Botanists. What each of these groups has in common is a method of communication tailored to their individual requirements.

In 1876, Alexander Graham Bell invented the telephone, possibly the greatest advance in language transmission since the invention of the alphabet. Bell, however, was not the first person in his family to garner international renown for a speech-related innovation. That accolade goes to his father, Alexander Melville Bell, who in 1864 devised a phonetic alphabet that he believed would enable deaf people to pronounce spoken language more accurately. He called it Visible Speech, because the character for each sound resembles the position that the throat, tongue, and lips must be in to make that sound, as shown opposite.

Alexander Melville Bell was an eminent Edinburgh elocutionist, married to a deaf woman, who devoted his life to the science of human vocalization. Upon inventing his alphabet, he moved the family to London to promote it. During demonstrations, his three sons would leave the room. A volunteer would recite words and phrases in challenging accents. Bell senior would write these words down in Visible Speech, and when the boys returned they would read his text and recreate the words and accents perfectly.

When two of his sons died of tuberculosis in quick succession, Alexander Melville Bell, his wife, and their surviving son Alexander Graham Bell emigrated to Canada, for the fresh air, and also to promote his alphabet in North America. The younger Bell's first job was teaching Visible Speech at the Boston School for Deaf Mutes. Even after his invention of the telephone made

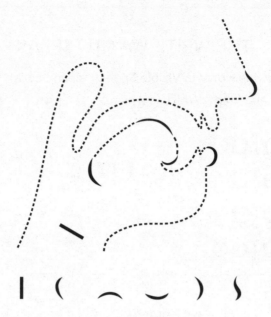

Each of the characters of Visible Speech are made up of combinations of the six symbols above, which denote positions in and around the mouth.

him a rich and celebrated technologist, Alexander Graham Bell never lost his commitment to Deaf education, founding a school for the Deaf (where he later met his wife, Mabel Hubbard). Visible Speech, however, never gained wide acceptance. It was quite complicated to learn, and an approach whereby deaf people were required to mimic the mouth shapes of the spoken word soon became outdated, losing favor to sign language.

In the following puzzle, think about how each character represents a mini-diagram of your mouth.

51

THE WRITE WAY TO SPEAK

The following words in Visible Speech mean "boot," "cogs,"
"peaks," and "tap." Which is which?

a)  ⊃ʃɑʊ

b)  ϴʃ⊃ʊ

c)  ⊃ʃ⊃

d)  ɑɟϴʊ

What are the following words?

e)  ϴʃɑ

f)  ⊃ʃʊ

g)  ʊɟϴ

h)  ⊃ʃʊ

Sir Isaac Pitman was the most prominent Victorian to argue that the Latin alphabet was not fit for purpose. In 1837, he published a system of shorthand that enabled users to write much more quickly. "Time saved is life gained," he preached. Pitman's teach-yourself book, *The Phonographic Teacher*, sold more than a million copies in his lifetime, and Pitman shorthand is still used today. (I learned it at journalism college and still use it, especially when I don't want anyone to read what I'm writing.)

Like Visible Speech, Pitman shorthand uses a phonetic alphabet in which each symbol stands for only one possible sound. With conventional spellings, the letters "ough" can be pronounced in at least half a dozen ways, as in the words "rough," "cough," "through," "thought," "though," and "bough." In Pitman's system, however, each symbol unambiguously represents a consonant or vowel sound, apart from those that represent a list of abbreviated common words. Using longhand, the average person can write about 30 words per minute (wpm). A competent writer of Pitman shorthand, however, will reach speeds of about 100 wpm.

If you want to go any faster, you'll need a stenotype machine.

Professional stenographers can transcribe speech at speeds of about 225 wpm, on average. That's almost four words a second—about as fast as you're reading this sentence. Stenotype machines use a keyboard layout (illustrated below) that has not changed since it was introduced in 1911. The machines are still widely used in the US, where there are about 30,000 stenographers. These speedy-fingered typists produce verbatim reports of court cases and government legislative proceedings. If you've ever watched a courtroom drama, or a debate on the Senate floor, you may have spotted the stenographer, usually close to the center of the room, sitting up straight, wearing an expressionless gaze while typing on a little machine. (In the UK, where criminal

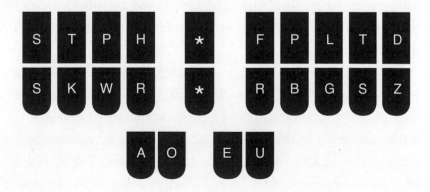

courts stopped using steno machines a few years ago, there are about 100 professional stenographers.) Stenographers are still used because, even though voice-recognition software has improved vastly in the last decade, a human stenographer is still more accurate than a computer at transcribing verbatim speech-to-text at speed. Humans are also much better at recognizing a change of speaker, and at putting in the right punctuation. As well as being present in courts, stenographers are used to provide communication support for deaf people, and live captioning at conferences and for TV.

The principle of stenography is that words are written syllable-by-syllable, based on how they sound, not how they are spelled. Each syllable is registered by pressing a combination of keys simultaneously, an action known as a "stroke." For example, a stenographer hearing the phrase "man bites dog" will press the machine three times: one stroke for "man," one stroke for "bites," and one stroke for "dog."

Strokes are made up of "chords," which are the keys pressed for individual sounds. The initial consonant of a syllable is always represented by a chord typed with one or more of the four fingers of the left hand, on the consonant keys to the left of the asterisk. The vowel sound is represented by a chord typed by the thumbs on the vowel keys. The final consonant of a syllable is represented by a combination of keys typed by one or more of the fingers of the right hand on the asterisk key and the keys to its right.

Let's look more closely at the stroke for "man." It consists of an initial consonant, "m," a vowel, "a," and a final consonant, "n." You will see that there is no M on the keyboard. When it is the intial consonant in the syllable, the sound "m" is represented by pressing the P and H on the left at the same time. The sound "a" in "man" is represented by an A. When it is the final consonant in the syllable, the sound "n" is represented by pressing the P and B on the right at the same time. So, a stenographer hearing "man" will simultaneously press P and H with the left hand, A with the left thumb, and P and B with the right hand. The output looks like this:

```
P  H  A              P  B
```

In some cases, such as the vowel sound "a" shown above, the letter in the printout is the same as its spelling. Often, however, as in the initial "m" and the final "n," the letters in the printout are not. Since the keys available to the left hand are different from those available to the right hand, each consonant sound has two possible chords: one when it is at the beginning of the syllable, and one when it is at the end.

The output of a steno text prints a stroke per line, so the output of "man bites dog" would be a "text" of three lines. The first line would have the stroke for "man," as shown above. The second line would have the stroke for "bites" (made up of the chord for "b" with the left hand, the chord for the vowel sound with the thumbs, and the chord for "ts" with the right hand), followed by a line with the stroke for "dog" ("d" with the left hand, "o" with the thumbs, and "g" with the right hand). If you make a mistake, the asterisk key functions as a backspace.

Before the computer age, when steno strokes were printed on folding paper tape at the back of the machine, the letters typed by any key appeared in a fixed position on the line. This meant that, whenever, say, the left P was typed, a "P" appeared in the same position that the left P appears in the example above. (I've kept that style for the following puzzle, even though computers now place the letters in order with no gaps, to save space.)

It takes a stenographer a few months to learn the basic lists of chords, and about two years of constant practice to get up to a professional level. Yet stenographers use other tricks to write as fast as possible: short forms and abbreviations. In the following puzzle, most of the lines in the output each represent a single syllable in the dialogue, but some are short forms and others are abbreviations.

52

FAST TALKING, FAST TYPING

The following dialogue took place in a courtroom.

The court: "Are you ready to enter a plea at this time?"
The defendant: "Yes, your honor."
The court: "How do you plead to counts one and two?"

The stenographer typed this conversation into a machine. The output, which was 25 lines long, was cut into the following five pieces, which have been arranged in random order.

(A)

```
                  O              P B
S T K P W H R            F R P B L G T S
          H       O     U
    T K           O     U
        P   H R A O   E                    D
```

(B)

```
S T K P W H R              F R P B L G T S
          R           U
          R       E
    T K         A O   E
    T           O
```

(C)

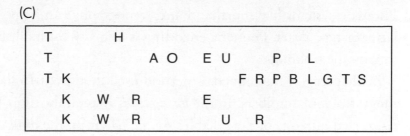

```
T           H
T               A O   E U     P   L
T K                     F R P B L G T S
    K   W   R           E
    K   W   R               U   R
```

(D)

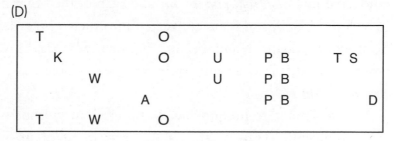

```
T                   O
    K               O       U       P B       T S
        W                   U       P B
            A                       P B               D
T       W           O
```

(E)

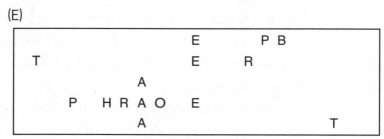

```
                    E           P B
    T               E       R
            A
        P   H R A O   E
            A                           T
```

1. Put the pieces into their correct order.

2. The next nine lines of transcription are shown below. What do they say?

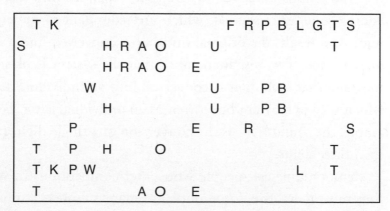

```
    T K                         F R P B L G T S
S               H R A O     U               T
                H R A O     E
            W               U       P B
            H               U       P B
        P                   E   R               S
    T   P   H   O                               T
    T K P W                             L   T
    T               A O   E
```

Phonetic systems like shorthand and stenography can help you write or type faster. Phonetic encoding is also a well-established mnemonic technique.

In chapter 3, I mentioned a method used in ancient India to memorize long numbers. In the *katapayadi* system, the digit 1 is converted into one of the syllables *ka, ta, pa,* or *ya*; the digit 2 is converted into *kha, tha, pha,* or *ra*; and each of the other digits 3, 4, 5, 6, 7, 8, 9, and 0 are converted into their own syllables. So if you needed to remember the number, say, 222,111, you would convert it into a memorable string of syllables, say, *kha-tha-pha-ka-ta-pa*.

A very similar technique has been used in Europe since the seventeenth century. The Major system takes its name from a self-help book published in London in 1845 by Major Bartłomiej Beniowski, an émigré Polish-Lithuanian cavalry officer—although he was only adapting a method that was, by then, already two centuries old.

In the Major system, every digit is converted into a sound. Each digit has one or more possible sounds that are unique to it. I'm not going to tell you the sounds that each digit converts to, since that is what you will deduce in the next problem. What I will reveal, however, is that once you've converted the digits into the sounds, these sounds are then expanded into words by adding some extra sounds (which are redundant to the encoding). As a result, the original number is converted into a word or sequence of words. Remember that the system is phonetic, meaning that each digit is converted into an individual sound, which may or may not be written as an individual letter. For example, the sound "tch" is the conversion of a single digit, rather than three digits.

Memory athletes—people who enter competitions in which they must memorize, for example, as many numbers as possible in a set time—still use the Major system. If you're looking for an impressive party trick, or want to memorize your credit card details, maybe you might want to learn it too. You are encouraged to convert the numbers into the silliest possible words, since the silliest words are the most memorable.

53
GREAT MEMORIES

Here are some examples of numbers converted into words using the Major system. (Note: The commas are not encoded; they are there to make the numbers easier to read.)

314,159,265,358	meteor tail banshee lime loaf
75,141,195	cluttered table
701,894	ghost vapor
99,501,247	pebble stone rock
7,512,026	golden snitch
8	hive
854	flower
2,394,429	number rainbow
6,096	cheese bush

Which numbers are expressed by:

a) moon cash   b) cat dog   c) gray elephant   d) short striped zebra

We first encountered the Cistercians in chapter 3. They are the monks belonging to the Catholic religious order who, between the thirteenth and fifteenth centuries, wrote dates in their own numerical notation. Another Cistercian curiosity is that, for almost a thousand years, they used their own sign language.

The story begins with the Benedictines, or Black Monks, so called because they wore black habits. In the tenth century, monks at one Benedictine abbey began a movement to return to a stricter ascetic life. Verbal communication was prohibited, and the monks began to develop their own sign language. In 1068 they drew up a list of the first 296 signs. Three decades later, one

A Benedictine monk (left) and a Cistercian monk (right).
Source: English Monastic Life, by Francis Adrian Gasquet, 1904.

of the monks from this abbey founded the Cistercian order, and maintained the silence rules. In contrast to the Benedictines, the Cistercians are known as the White Monks, because their habits are white.

The pictures in the next problem are all from a 1975 book by Robert Barakat, *Cistercian Sign Language*. The monks are from St. Joseph's Abbey, near Worcester, Massachusetts. The author notes that "since brief verbal communication was permitted about five years ago, the use of signs among the brothers in some monasteries has decreased greatly." His worries in 1975 proved well-founded: the thousand-year tradition has died out as a working means of communication, although there are still some signers. Father Joseph, of Mount St. Bernard Abbey, the only Cistercian abbey in England, told me (via email): "Some of the most elderly monks can remember it, but it is no longer taught to newcomers."

In Cistercian Sign Language, words are broken down into simpler concepts. For example, the word "beer" is expressed by making the sign for "corn," followed by the sign for "water." What makes the following problem fun is that many of the signs were devised to resemble the concepts they are describing.

54

A GODLY SILENCE

Shown below are 16 words in Cistercian Sign Language and their translations in English, listed in alphabetical order. Each individual sign is labeled with a number. Match the words to their correct translations.

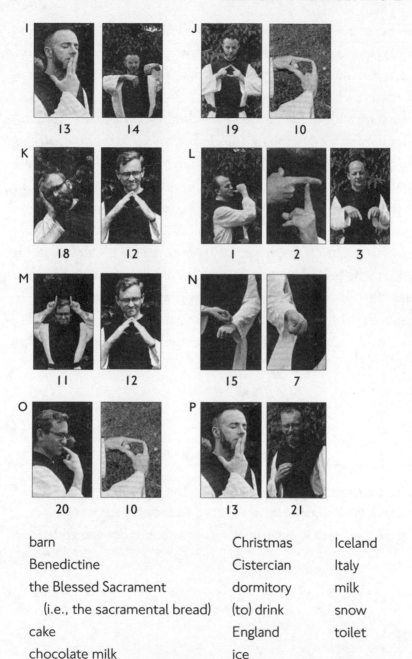

barn

Benedictine

the Blessed Sacrament
  (i.e., the sacramental bread)

cake

chocolate milk

Christmas

Cistercian

dormitory

(to) drink

England

ice

Iceland

Italy

milk

snow

toilet

HINT: The pictures include individual signs for "God,"
"black," and "white," each of which is used twice. One sign
represents a consonant, and one sign a vowel.

The next three problems concern tactile alphabets for blind people that use grids of embossed dots. The task in each will be to discover the ordering principle that lies behind the choice of dot patterns.

We start with Braille, the most widespread system in use. Louis Braille was a Frenchman, blinded in an accident at age three, who devised his alphabet in 1824 when he was only 15 years old. A pupil at the world's first special school for the blind in Paris, he got the idea from a French military code of dots and dashes impressed onto thick paper that enabled soldiers to communicate silently at night. Braille never saw the success of his alphabet: He died, at 43, before his school had even adopted it. A hundred years after his death, however, he was such a famous figure in France that his remains were exhumed from his home village (minus the hands, which stayed) and reburied at the Panthéon in Paris, the resting place of national heroes.

55

PIZZAS AND VERMOUTH

Here are three sentences in Braille. The large black dots represent embossed letters. I have added smaller dots to show how each letter is built within a six-dot domino grid.

This fox is too quick!

How old are you, Jane?

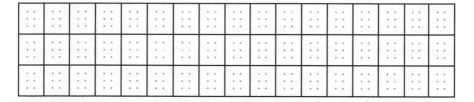

She is 89 years old.

Write out in Braille:

Bring 40 pizzas and vermouth, Mark!

To complete this problem, you will need to fill in the table below. You will see that I have put "w" in the bottom-right corner. That's because, in French spelling, the "w" almost never appears.

a	b	c	d	e	f	g	h	i	j
k	l	m	n	o	p	q	r	s	t
u	v	x	y	z					w

The US developed its own tactile alphabets for the blind. At the beginning of the twentieth century, the most widely used system was New York Point, invented by the educator William Bell Wait at the New York Institute for the Education of the Blind. Yet many prominent blind people, such as the author and activist Helen Keller, preferred Braille. A series of public hearings took place in the 1910s to decide which system would become the national standard. Such was the passion on both sides that the battle for alphabet supremacy became known as the War of the Dots.

The main argument in favor of New York Point was its brevity. It uses fewer dots overall than Braille does. As we saw above, every letter in Braille is set on a domino grid, three dots high and two dots wide. In New York Point, however, the letters are two dots high, and can be anything from one to four dots wide.

The nonsense phrase "etaoin shrdlu" is a list of the most common letters in the English language in approximate order of frequency. In New York Point, the "e" and the "t" are each made up of a single dot. The next five letters in the sequence are made up of two dots over two columns. The less commonly used letters use three columns, and capital letters four. (There are also separate dot patterns for common English letter combinations like "ing" and "ch.") As a result, about half of all the letters in a New York Point text will be set in a grid that's only two dots high and either one or two dots wide, and they will use either one or two dots. Advocates of New York Point argued that, thanks to this efficiency, books would be shorter and cheaper to produce.

The War of the Dots was won by Braille, for a number of reasons, including its better system for punctuation and the fact that it could be read faster. New York Point, once the toast of American blind education, was condemned to obscurity (and is brought back here for one night only).

56

I • NEW YORK

Match the names Ashley, Barb, Carl, Dave, Elena, Fred, Gerald, Heather, Ivan, Jack, Kathy, and Lisa to their equivalents in New York Point.

a.

g.

b.

h.

c.

i.

d.

j.

e.

k.

f.

l.

Write "Orson" in New York Point in the grid below.
(You will not need to use all the columns for each letter.)

Several languages that do not use the Latin alphabet have their own versions of Braille. The Japanese system, called *tenji*, is fantastically concise. Each six-dot grid encodes a syllable, comprising an initial consonant and a vowel, or a vowel on its own.

57

DOTTY ABOUT JAPAN

Here is a word written in Japanese Braille. The thick dots are raised bumps, and the tiny dots are empty positions.

karaoke

Here are six more Japanese words. Their pronunciations are listed in a different order. Match the symbols to the correct words.

a. *atari*

b. *koi*

c. *haiku*

d. *katana*

e. *kimono*

f. *sake*

What are the following words?

g.

h.

People who are blind from birth usually find Braille the most appropriate tactile writing system. Yet those who lose their sight later in life, and those with a less keen sense of touch, often find it easier to learn the Moon alphabet, which contains embossed letters that bear a resemblance to their Latin equivalents.

The Moon alphabet was the first widely used tactile system in the UK. Its inventor, William Moon (below), lost his sight at age 21 after contracting scarlet fever. He devised his system in the 1840s, and a decade later set up a printing press in Brighton. Books in Moon's type were long: One of the first books he printed was the Bible, which required 4,000 pages over 58 volumes. Moon traveled internationally promoting his system, and was supported by missionary bodies, who introduced it around the world. It is said that his printing press published material in 400 languages.

Moon type is still read by a small number of people in the UK, and the Royal National Institute for the Blind keeps a library of Moon books for loan.

William Moon, taken from his 1877 book *Light for the Blind.*

58

TICKET TO THE MOON

Each phrase below is written using both the Moon alphabet and the Latin alphabet. The list on the left is not in the same order as the list on the right. You will need to deduce some of the Moon letters by their similarities to their Latin versions.

Moon	Latin
∩ˉ\ˉ	we know
ı ∟ ⌐	loneliness
∟⁻:/	gravity
⊂⊓:−	Manchester
∟ʌ<ˉ	Whollowitz and I
∩ˉ <	commenting
∠ʌ∩⁻ˉ\	I like you
ı∩ʌ\ı⊃ ı/ ʌ ᴄı∟⊃	10 tickets
⁻ı /−\ʌ∩∪ˉ\\ıˉ/	where
⌐ o ⁻ <ˉ⌐	I am from Quasar
∩	knowing
∩\ʌ∨:⌐	Ingrid is a child
⁻∪ ∟\/	lake
∟O∩ˉ∟ı:/	panther
<⁻:∩	Xerxes in the jail
⊓ʌ∩ᴄˉ/−ˉ\	you have the key
⊂⊓:−:∩	Likeness
∩O∟∟O∩ı−z s ı	2 letters
⁻ʌ⌐ −ıᴄ<ˉ−/	9 strawberries
>ˉ\>ˉ/ ı∩ ⁻ ⌡ʌı∟	go
ı ʌ⊓ ⌐ ⊔ʌ/ʌ\	comment

Fill in the grid below with the correct Moon character for each letter. You may be wondering why the English list contains such strange words and phrases. It is to supply you with the letters w, q, x, and z. Some combinations of letters, such as "ch," have a single character, and some suffixes, such as "-ing" and "-ness," have special symbols. There are also some words that use abbreviations.

A		F		K		P		U		Z		-ING	
B		G		L		Q		V				-MENT	
C		H		M		R		W		CH		-ITY	
D		I		N		S		X		TH		-NESS	
E		J		O		T		Y		WH			

Translate the following text:

Γ ⁚ ᴐUΛ∕Λ∖ CΛ⊓Γ ⁚ ∩IZΛ∖ᴐ ϾILᴐ∖ΓN..

..Γ∖U−UΓ ⁚ ∕−CIᴐΓ∖ᴢ ϾI∩ ∖L Λ ο ⅃Γ⁚

⁻Λᴐᴐ ΓI∖Γ ∕−O∖⁊∕ ∩ILL ∕οOC< CI−IΓ∕ S

NΓ⁚ ..Γᴢ∧C∕Γ LLI∩ I S ⅃ −Uᴗ .∕ᴐLΓIΓ

⁚ ZΛ∖OVO ϾILᴐ ∩ILL ᴗΓ <Γᴢ− ∩I⁚ ⁊Λ>I

..∖ΓVΓ∖oΓ

Color blindness—the inability to distinguish colors—affects about 1 in 12 men, and 1 in 200 women. It can be a serious disability. People with total color blindness struggle with many everyday tasks, from reading traffic lights to choosing the right item of clothing, and from telling the difference between ripe and moldy food to following the right path on a map.

The Portuguese graphic designer Miguel Neiva devised a labeling system for color-blind people using black and white symbols to represent colors. The code, ColorADD, is now widely used in Portugal, where you will find it on colored pencils, clothes, subway maps, medicines, indoor parking levels, recycling bins, and traffic lights. ColorADD uses six basic shapes, and can describe 50 colors, from navy blue to gold and from khaki to light green.

THE COLOR PURPLE

Here are four ColorADD symbols with the colors they represent.

orange   green   dark blue   gold

1. Draw the symbols for red, yellow, and brown.

2. The four colors below are dark purple, pink, silver, and white, but not in that order. Match the colors to the symbols.

3. What are the two possible symbols for gray?

All this talk of purples and pinks is making me think of flowers.

Botanists write the structure of flowers in two ways: with floral diagrams, and with floral formulae—ideas that were both developed in the nineteenth century. By convention, K refers to sepals (the outer leaves of a flower), C to petals, A to stamens (the male, pollen-producing stalks), and G to carpels (the female parts that develop into fruit on fertilization). The rules for the formulae have been simplified for the purposes of this puzzle.

60

BUDDING BOTANISTS

Here are some floral diagrams and their floral formulae:

$$K_4C_4A_4G_1 \qquad K_0C_5A_{10}G_1 \qquad K_{(4)+1}C_5A_{4+1}G_5 \qquad K_4C_{(2+2)}A_{4+2+2}G_4$$

Write the formulae for the following three diagrams:

a. b.  c.

LINGO BINGO

Vocab Test M–Z

Guess the correct meaning of the word.

1. MAMMOTHREPT
   a) Suckling
   b) Mythical chimera: part mammoth, part reptile
   c) Stolen baby
   d) A spoiled child

2. MOFETTE
   a) Taffeta-like material
   b) Vapors from a volcano
   c) Fluffy person
   d) Preliminary sketch

3. PACHANGA
   a) Obsolete horse-drawn tram in South America
   b) Cashier in Peru
   c) Bolivian expletive
   d) Dance music of Cuban origin

4. PELHAM
   a) Horse's bit
   b) Light overcoat
   c) Town carriage
   d) Variety of pear

5. RHODOLOGY
   a) Study of Rhodes
   b) Study of roses
   c) Badge collecting
   d) Use of rhodium to catalyze slow reactions

6. SPARTINA
   a) Tiny Spartan woman
   b) Mast varnish
   c) Type of grass
   d) Derisive term for an upstart state

7. TERCEL
   a) Small animal
   b) Male falcon
   c) Tithe, as it were, of a third
   d) Fraction in base 3

8. TERPODION
   a) Three-legged dinosaur
   b) Alien from *Doctor Who*
   c) Stool
   d) Keyboard instrument that produces sound by friction

9. TOISON
   a) Lamb's wool
   b) Rallying cry
   c) Medieval French love song
   d) A point in real tennis

10. WIGAN
    a) Rubberized cotton fabric
    b) Slang for a judge
    c) Mythical land
    d) Out-talk

11. WISENT
    a) Smart-ass
    b) European bison
    c) Go-between
    d) Saxon emissary

12. WITZELSUCHT
    a) Addiction to telling jokes
    b) Brazilian footballer
    c) Finnish doggedness and grit
    d) Firewalking

13. YAWL
    a) Shout incomprehensibly
    b) The angle or degree of yaw of a ship
    c) Two-masted sailing boat
    d) The second derivative of the jerk in calculus

14. ZOPISSA
    a) The intercept on the imaginary axis
    b) One who pretends to like wild animals
    c) Ointment made from scrapings from the sides of
       ships
    d) Parasite of a parasite

7

Treacherous Subjects

ᴚ

LANGUAGES
ON THE EDGE

Every language has its own idiosyncrasies. This is especially true of English: We've already seen how our spelling makes us an international laughingstock, but that's only the start of our deviance, isn't it? For example, that phrase I just used—"isn't it?"—is a grammatical structure of such extraordinary complexity that almost no one learning English as a second language masters it properly. And the fact that we change our word order when asking questions—"This is hot," but "Is this hot?"—is a completely inexplicable thing to do, according to the speakers of almost 99 percent of the world's other languages.

This chapter is a showcase of languages with particularly curious features of their own. I've chosen phenomena that differ as much from English as I could find. You will be asked to spot things you don't even know you are looking for, which makes some of the questions hard—but especially rewarding when you learn the answers. Inevitably, this chapter is the most technical in the book. I'll be talking about parts of speech, word order, number, gender, and other basic concepts of linguistic analysis. You will be baffled and amazed. The diversity of languages in the world is fascinating and remarkable.

But before I bemuse you with exotic tongues, here's a reminder of how perplexing English grammar can be.

(Some basic terminology: A "verb" is a word that expresses an action. The "subject" of a sentence is the do-er of the action and the "object" is the do-ee.)

61

WEASEL WORDS

The following sentence, though bizarre and deliberately confusing, is grammatically correct:

> The weasel that a boy that startles the cat thinks loves smiles eats.

Answer the following questions. In some cases, the answers may be "nobody/nothing in this sentence."

1. What is the subject of this sentence?
   (Give a single-word answer.)
2. How many verbs are in the sentence?
3. Who startles whom or what?
4. Who thinks what?
5. Who loves whom or what?
6. Who smiles?
7. Who eats whom or what?

Tajikistan is a mountainous country of about ten million people in Central Asia. Its inhabitants mostly speak Tajik, an Indo-European language closely related to Farsi. Tajik has been written in the Cyrillic alphabet since 1939, when Tajikistan was part of the Soviet Union. (Before the Russians invaded, it used the Perso-Arabic alphabet.)

62

GOOD NEIGHBORS AND GOOD FRIENDS

Here are three Tajik phrases and their translations:

дӯсти хуби ҳамсояи шумо	a good friend of your neighbor
ҳамсояи дӯсти хуби шумо	a neighbor of your good friend
ҳамсояи хуби дӯсти шумо	a good neighbor of your friend

What are the Tajik words for "friend," "good," "neighbor," and "your"?

The Caucasus region.

Political upheaval on the fringes of the Soviet empire created even more alphabetic disruption in Abkhazia, a small autonomous part of Georgia on the Black Sea coast. Between 1892 and 1954, it used eight different alphabets (three types of Cyrillic, three Georgian, and two Latin), although this chaos was partly due to the peculiarities of Abkhaz, the language that is spoken there. All language lovers should be aware of Abkhaz, if only because it is the only word in English that begins in an "a" and ends in a "z" that is not a place name. Yet what really makes Abkhaz stand out is its remarkable number of consonants, about 60, compared to its number of vowels, which is two. (English has about 25 consonants and about 20 vowels, depending on accent.) In fact, the ratio of consonants to vowels in Abkhaz is one of the highest of any language in the world, if not the highest. Linguists salivate over its unusual consonant sounds, such as "тә," which is a "t" and a "p" spoken at the same time, and its tongue-twisters, such as ҧстәкәоуп—which means "they are animals"—a single syllable that starts with a cluster of four consecutive consonants.

In Abkhaz, all nouns and adjectives begin with the prefix *a-*, as you will see in the following problem. Gotta get them vowels in somewhere! The problem also reveals several other linguistic features of Abkhaz that are different from English. I'll outline the relevant ones for you now.

Word order in English is subject-verb-object, usually abbreviated to SVO. For example, in the sentence "The mothers are wearing the trousers," the subject ("the mothers") comes first, followed by the verb group ("are wearing") and then the object ("the trousers"). In the problem below, each of the Abkhaz sentences consists of three words: a subject, a verb, and an object. You'll quickly realize that Abkhaz word order is not SVO. Once you have figured out what it is, you will be able to work out the English translations of each Abkhaz word. Well done. You have now completed the easy part.

To solve the problem fully, however, you need to deduce three more rules about Abkhaz. One concerns how to make the plural form of a noun. Another concerns how the beginning of the verb changes depending on the subject. And the third concerns why these sentences use two forms of the word "wear."

(I've simplified the English transcription to remove many of the phonetic symbols that Abkhaz transliteration inevitably requires, although some remain. You do not need to know how these symbols are pronounced to solve the problem. If you are interested, however, see the Appendix.)

THE MOTHERS ARE WEARING THE TROUSERS

Here are some Abkhaz sentences in simplified transcription, with their English translations. (A *cherkeska* is a traditional Caucasian piece of clothing: a single-breasted collarless coat.)

Anchwa apərahwa ashwup'	The god is wearing the apron
Anchwa ajkwa rəshop'	The mothers are wearing the trousers
Aesh axəlpa ashwup'	The squirrel is wearing the hat
An ak'wəmʒwə Ishwup'	The mother is wearing the cherkeska
Atahwmadachwa ac'atɛ'kwa rshwup'	The old men are wearing the coats
Abachwa ak'wəmʒwəkwa rshwup'	The sons are wearing the cherkeskas

Adzʁab ajmsəksha ləshop'	The girl is wearing the felt boots
Ab ajkwa ashop'	The billy goat is wearing the trousers
Atahwmada apərahwa ishwup'	The old man is wearing the apron
Abkwa ak'wəmʒwəkwa rshwup'	The billy goats are wearing the cherkeskas
Aeshkwa ak'asəkwa rshwup'	The squirrels are wearing the shawls

Choose the correct Abkhaz translation for each of these English sentences:

1. The son is wearing the trousers
    a) *Ab ajkwa ashwp'*
    b) *Ab ajkwa ishwup'*
    c) *Aba ajkwa ashop'*
    d) *Aba ajkwa ishop'*

2. The girls are wearing the shoes
    a) *Adzʁabchwa ajmaakwa rshwup'*
    b) *Adzʁabchwa ajmaakwa rəshop'*
    c) *Adzʁabkwa ajmaakwa rshwup'*
    d) *Adzʁabkwa ajmaakwa rəshop'*

A prime contender for the title of the world's most unusual and complicated language is Navajo, the Native American language spoken by about 170,000 people in the American southwest. Its grammar is so unfathomable that the US military used it, with great success, as a code during the Second World War.

The idea that a Native American language could be used to encrypt military secrets had emerged in the previous world war, when, in 1918, a US army officer in the Allied trenches overheard two soldiers chatting in Choctaw (which now has about 10,000 speakers in Mississippi and Oklahoma). The officer realized instantly that the soldiers were perfect encryption devices, since the Germans would be as flummoxed by the language as he was. Within hours, a squad of eight Choctaw soldiers were dispatched to strategic positions. These "code talkers" passed information to each other in Choctaw without the Germans being able to decipher it. (According to one story, German wiretappers thought the US had invented a way of speaking underwater.) Using Choctaw was much faster than using a machine to encode and then decode a message, although it was not perfect, since Choctaw did not have words for relevant military jargon. For example, the Choctaw used the phrase "little gun shoot fast" to mean "machine gun," the ambiguities of which were open to misinterpretation.

For the Second World War, the Americans had more time to plan their system of code talkers. This time they chose to use Navajo since, in the words of the official report, it "is completely unintelligible to all other tribes and all other people, with the possible exception of as many as 28 Americans who have made a study of the dialect." The report ruled out languages spoken by other Native American communities because it claimed that they had all been "infested" with German linguistics students over the previous two decades.

The US military employed 420 Navajo as code talkers. Each of them memorized a lexicon in which common Navajo words had specific military meanings. The word for "hummingbird," *da-he-tih-hi*, for example, meant "fighter plane." Military historians believe that the Americans would never have captured Iwo Jima, one of the significant victories of the war in the Pacific, without the Navajo code talkers, who sent and received hundreds of messages, all without error and none of which were deciphered by the Japanese. According to Simon Singh, author of *The Code Book*, Navajo remains one of the very few codes in history that was never broken.

Navajo makes for a fearsome code because it has many features unknown in Asian or European languages. One of them relates to word order, and is the basis of the next problem. Navajo word order can be subject-object-verb (SOV) or object-subject-verb (OSV). What is particularly curious, however, is that depending on the meanings of the words, sometimes both word orders are grammatical, and sometimes only one of them is. Warning: This problem requires military-level code-breaking skills. Whether or not you get the correct answer, you will find the solution fascinating.

(This question contains many unusual phonetic symbols. As before, you don't need to know the pronunciations in order to solve the problem. If you are interested in how they sound, however, see the Appendix.)

64

GAH! THE GOPHER KILLED THE ANT

Here are some sentences in Navajo and their English translations. Each pair of Navajo sentences has the same meaning; note, however, that sentences marked with an asterisk (*) are ungrammatical. (A gopher is a rodent about the size of a hamster.)

Ashkii diné biztał.
Diné ashkii yiztał.
} The man kicked the boy.

Ashkii diné yiztał.
Diné ashkii biztał.
} The boy kicked the man.

Ashkii łééchąąʔí yiztał.
*Łééchąąʔí ashkii biztał.
} The boy kicked the dog.

Awééchíʔí mósí biztał.
*Mósí awééchíʔí yiztał.
} The cat kicked the baby.

*Awééchíʔí diné yiztał.
Diné awééchíʔí biztał.
} The baby kicked the man.

Awééchíʔí shash binoołchééł.
Shash awééchíʔí yinoołchééł.
} The bear is chasing the baby.

Mósí naʔazízí yinoołchééł.
*Naʔazízí mósí binoołchééł.
} The cat is chasing the gopher.

*Mósí shash bishxash.
Shash mósí yishxash.
} The bear bit the cat.

Naʔastsʔǫǫsí tsísʔná bishish. *Tsísʔná naʔastsʔǫǫsí yishish.	The bee stung the mouse.
Naʔazísí wóláchíí yiisxí. *Wóláchíí naʔazísí biisxí.	The gopher killed the ant.
Tsísʔná naʔashjéʔii yishish. Naʔashjéʔii tsísʔná bishish.	The bee stung the spider.
Naʔastsʔǫǫsí naʔazísí yishxash. Naʔazísí naʔastsʔǫǫsí bishxash.	The mouse bit the gopher.

Translate the following four sentences into English, and indicate whether or not they are grammatical:

1. Awééchíʔí łééchąąʔí binoołchééł.
2. Tsísʔná ashkii bishish.
3. Naʔastsʔǫǫsí naʔashjéʔii bishxash.
4. Wóláchíí diné yiisxí.

The following two sentences are grammatical. Suggest a meaning for the word *gah*.

Gah mósí biisxí.
Mósí gah yiisxí.

The Kwakwa̱ka̱'wakw people live on the Pacific coast of Canada, on the northeastern coast of Vancouver Island, and across the water on the mainland. They are well-known in Canada for their elaborate artwork, including masks and totem poles, and for their *potlach* feast ceremony. About 4,000 people claim Kwakwa̱ka̱'wakw ancestry, and many of them are fishermen. The local delicacy is food dipped in oil from the *eulachon* (pronounced "ooligan"), or candlefish, which gets its name because the dried fish is so oily it can be burned like a candle.

The Kwakwa̱ka̱'wakw speak Kwak'wala, which has several features—two of which are highlighted in the next puzzle—that make it unlike most of the world's major languages. First, it has more than 40 consonant sounds and only six vowel sounds, giving it a very high consonant-to-vowel ratio (although not as extreme as that in Abkhaz). The differences between some of the consonant sounds are barely perceptible to outsiders, even to linguists who have been studying the language for years. The standard orthography considers the two-character sequences *tl*, *dl*, *ts*, *dz*, *dw*, *gw*, and *xw* as single consonants. Underlining a letter means that the speaker should press their tongue further back in the mouth—so *a̱* is pronounced "uh." And the apostrophe indicates increased pressure at the back of the mouth.

The second feature of Kwak'wala that earns it a place here is its interesting morphology. The language has a small set of roots, and words are created by adding suffixes to these roots. In the next puzzle, for example, you'll see there is a root for "write," and that four new words are created by adding suffixes: one that gives you the verb "to write," another that gives you the instrument that does the writing, and so on. Let's say that "X" is a root. Kwak'wala has a suffix that means "to X on the forehead," one that means "to X while standing in shallow water," and one that means "to X in a dream." In fact, the diversity of meaning of Kwak'wala's suffixes—of which there are about 400—is probably unique in world languages.

When you learn Kwak'wala, you don't learn a set of fixed words. Instead, you learn the roots, and the rules for building new words from these roots (such as changing the final consonant of the root, for example). "It's very creative. Kwak'wala speakers are constantly making up new words, and this is a crucial part of the grammar," says linguist Katie Sardinha, who did her PhD on the language.

Only about 150 people speak Kwak'wala as their native tongue, and most of them are over 65, which makes it severely endangered. Some of these elders now go into schools to speak to young Kwakwaka'wakw, so there is a glimmer of hope that, even as the number of fluent speakers declines, some understanding of the language will survive.

The Kwakwaka'wakw live in the Vancouver area.
(The other landmarks relate to puzzle 81.)

65

THE SNOB CONTAINER

The following word search contains 30 words in Kwak'wala. Fifteen of them are already identified, and are listed opposite, together with their meanings. Find the remaining 15 words, and match them to their meanings, which are also listed opposite.

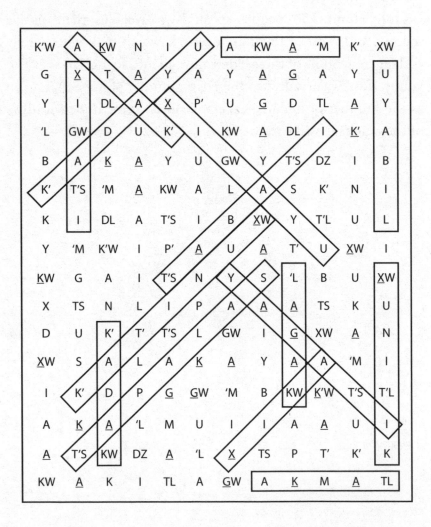

to write	K'	A	T	A			
pen/pencil	K'	A	D	A	Y	U	
something written	K'	A	D	A̲	KW		
writer	K'	A	T'	I	N	U	X̲W

to sweep	X̲	I	KW	A		
broom	X̲	I	GW	A	Y	U
dustpan	X̲	I	GW	A	T'S	I

| bowl for candlefish oil | T'S | A̲ | B | A | T'S | I |
| food for dipping in oil | T'S | A̲ | P | A | L | A | S |

to be proud/to be a snob	TL	A̲	M	K̲	A	
to iron something	'M	A̲	KW	A		
berry cakes	'L	A	G̲	A̲	KW	
deck of cards	L	I	B	A	Y	U
fisherman	K	I	T'L	I	N	U	X̲W
knitting basket	Y	A̲	G̲	A	T'S	I

to catch fish with a net

to dip food in candlefish oil

envelope

expert card player

expert knitter

fishing boat

(an) iron

to knit

knitting needles

to make berry cakes

to play cards

something knitted (like a
    sweater)

tourist boat/cruise ship/
    ferry

wool

wrinkled clothes

What is the world's weirdest language? In 2013, Tyler Schnoebelen, a computer scientist, decided to find out. He ran a statistical analysis of the *World Atlas of Language Structures* (WALS), a database of the linguistic properties of almost 3,000 languages kept by the Max Planck Institute for the Science of Human History, in Germany. He singled out 21 distinct linguistic features, such as word order, phonology, and so on, and gave each language marks for how unusual it was for each feature. He restricted himself to languages that had enough data across the features he was looking at, which reduced his list to 239. His "weirdness" ranking put English in thirty-third place and German in tenth.

The top three, in reverse order, were:

3. Choctaw, which I've already mentioned.

2. Nenets, which has about 20,000 speakers in Siberia, Russia.

1. Chalcatongo Mixtec, which has about 6,000 speakers in Oaxaca, Mexico. An insight into its "weirdness" is provided by the next puzzle.

66

THE TRUTH ABOUT MIXTEC

Here's a sentence in Chalcatongo Mixtec, with its English translation. The language has been slightly simplified for the question. (As previously, the symbols and accents are not relevant for solving the problem, but you can find out how to pronounce them in the Appendix.)

Nduča kaa ñíʔní   <u>The</u> water is hot

The underline means that the word is emphasized. We are talking about a particular volume of water.

Here are seven more sentences, with the translations in random order. Match the sentences to their correct translations.

1. *Maria kúu ǂ xasɨʔɨ*
2. *Ñíʔní nduča*
3. *Juan kaa lúlí*
4. *Ndežu kaa žaʔu*
5. *Sɨʔɨ Maria*
6. *Juan kúu xažiirí*
7. *Pedro kúu xalúlírí*

a) Pedro is my child
b) Juan is my husband
c) Maria is a woman
d) The water is hot
e) Maria is feminine
f) <u>The</u> food is expensive
g) Juan is small/short

Here are four more Chalcatongo Mixtec words:

ndáa: true   *kǔnú*: deep   *kwaʔá*: red   *saʔma*: clothes

Translate the following into Chalcatongo Mixtec. (The last three translations contain no word for "it.")

i) depth
ii) The clothes (unstressed) are red
iii) My clothes are the red ones

iv) <u>It</u> is true
v) It is true
vi) It is the truth

Traveling southward, I thought I'd raise the tone—and the altitude—with a poem from the Andes. Quechua is spoken by about 10 million people in Peru, Ecuador, and Bolivia, making it the indigenous American language with the most speakers. It is an "agglutinative" language, which means that words are created by combining distinct elements together, a bit like threading a sequence of beads on a string. For example, the phrase "in my house" in Quechua is a single word *wasiypi*, which is made from *wasi*, the word for "house," plus *-y*, the suffix for "my," plus *-pi*, the suffix for "in." Thanks to agglutination, a single word in Quechua can often express what would take several words in English. In the poem below, for example, each line of English translation contains roughly twice the number of words as the Quechua.

67

DARKEST PERU

The following eight couplets are from a poem by the twentieth-century Peruvian poet Sisku Apu Rimac ("Sisco who talks to the spirits"). The original Quechua couplets are listed in random order in the right-hand column. (The *vicunya* is a wild relative of the alpaca, and *kule* and *puku* are birds.)

1. For what, God,
   Did you create my suffering?

   a) *Kule kuleq thapanpichus*
   Taytallayri churyawarqa

2. Did you never know
   What happiness is?

   b) *Kunan kuna waqanaypaq*
   Urqun qasan purinaypaq

3. Maybe in the nest of the pukus
   My dear mother gave birth to me.

   c) *Wikunyachus mamay karqa*
   Tarukachus taytay karqa

4. Maybe in the cradle of the kules
   My dear father engendered me.

d) *Manataqchu yacharqanki*
   Imaynas kawka kayta

5. Like the poor puku
   I endure the cold winds.

e) *Imapaqmi Apu Tayta*
   Nyak'ariyta kamarqanki

6. Or the poor kule
   I cry as I suffer.

f) *Puku unya hina*
   Chiri wayra muchunaypaq

7. Perhaps my mother was a vicunya;
   Perhaps my father was a deer;

g) *Puku pukuq qesanpichus*
   Mamallayri wachawarqa

8. And for these reasons I cry while
   I wander through the highlands.

h) *Kule unya kaqlla*
   Nyak'arispa waqanaypaq

a) **Match the couplets to their correct translations.**
b) **Write down the Quechua words for "poor," "suffer,"
   "mother," and "deer."**
c) **What is the literal meaning of the Quechua word for
   "God"?**

Papua New Guinea is the most linguistically diverse place on Earth, as I mentioned in an earlier chapter, so you might expect it to have languages with extremely unusual linguistic phenomena. One example is Walman, which is spoken by about 1,700 people in four villages on the country's north coast. In Walman, the word for "and" behaves like a verb, taking different forms depending on the words that come before and after it.

68

"AND" AND "AND"

Below on the left is a list of English phrases that include the word "and." On the right is a list of terms that represent how the "and" would be translated into Walman.

me and you	mcha
me and my brother	man
you and the chieftain	nan
you and your neighbors	nay
the neighbor and I	npa
the chieftain and his wife	na
the neighbor's mother and I	wpa
my wife and the guests	way
your sisters and you	ycha
my brothers and the chieftain's mother	ya

What is the correct Walman form of "and" in the following phrases?

1. the chieftain and you
2. your neighbors and I
3. my brother and the chieftain
4. grandmother and my wife
5. the guest and the hosts
6. you and the guests' wives
7. me and the chieftains
8. the neighbors and the guests

About one hundred miles east of the Walman-speaking villages is the Sepik River, one of the world's largest rivers in terms of water flow. The people who live on the banks of the Sepik, such as the Iatmül, face regular floods. The 47,000 speakers of Iatmül have only recently had contact with Western civilization. Exposure to new ideas can lead to words gaining unexpected meanings.

A CANOE PROBLEM

Here are some words in Iatmül and their English translations:

Iatmül	Literal translation	Meaning
vi	spear	spear
ni'bu	land	land
guna vaala	water canoe	canoe
walini'bana gu	white person's water	soda/alcohol

What is the meaning of:

> walini'bana vi
> ni'buna vaala

Match the following three words to their correct meanings. You will need to think about the order in which the words entered the Iatmül language.

laavu	to read
laavu-ga	banana
laavu-ga vi'	book

(Note: The word vi' is unrelated to vi.)

We finish this chapter in Africa. The world's largest language family, in terms of the number of languages it contains, is Niger-Congo, which is made up of about 1,500 languages spoken across sub-Saharan Africa. A linguistic feature shared by almost all the branches of Niger-Congo is a system of "noun classes," in which nouns are assigned to one of several categories that determine how they behave. Noun classes function a bit like gender does in languages with gender. Whereas French has two genders, however, and German three, Niger-Congo languages can have more than a dozen noun classes.

Swahili is the most spoken language in Africa. It is the lingua franca of eastern Africa, spoken across Tanzania, Kenya, and Uganda, mostly as a second language.

70

BAD CHILDREN HAVE SMALL UMBRELLAS

The following Swahili sentences are translated into English. Swahili has no words for "the" or "a."

Mtu ana watoto	The man has children
Watu wana mifuko	The men have bags
Mtoto ana kiazi	A child has a potato
Viazi vibaya vinatosha	The bad potatoes are enough
Kijiko kikubwa kinatosha	A large spoon is enough
Mtoto ana mwavuli	The child has an umbrella

Mto una visiwa vikubwa	The river has large islands
Mwavuli una mfuko mdogo	The umbrella has a small bag
Wafalme wana vijiko vidogo	The kings have small spoons
Kisiwa kikubwa kina mfalme mbaya	The large island has a bad king
Mito mizuri inatosha	Good rivers are enough
Watoto wabaya wana miwavuli midogo	Bad children have small umbrellas

Translate the following into Swahili:

1. The small children have bad spoons

2. A large umbrella is enough

3. The large islands have rivers

All languages have something curious about them. The examples in this chapter only skim the surface of the vast linguistic diversity in the world. If there was ever a language that had nothing weird about it, now *that* would be really strange.

LINGO ⓑⓘⓝⓖⓞ

Special Letters

English is unique among European languages using the Latin alphabet in being almost entirely free of diacritics, that is, adornments such as accents, cedillas, and circumflexes that indicate how a letter is to be pronounced.

This question presents 12 translations of the statement "All human beings are born free and equal in dignity and rights," the first line of the Universal Declaration of Human Rights. Can you guess the language?

If the translation contains a diacritic that appears only in that language, I have marked it. The other languages are recognizable by their unique combinations of consonants and vowels.

1. *Minden emberi lény szabadon születik és egyenlő méltósága és joga van.*

   (Unique letter: ő)

   a) Hungarian
   b) Polish
   c) Slovak
   d) Turkish

2. *Tots els éssers humans neixen lliures i iguals en dignitat i en drets.*

   a) Basque
   b) Catalan
   c) Corsican
   d) Maltese

3. *Øll menniskju eru fødd fræls og jøvn til virðingar og mannarættindi.*

   (Only language that uses both ð and ø)

   a) Danish

   b) Faroese

   c) Norwegian

   d) Swedish

4. *Gizon-emakume guztiak aske jaiotzen dira, duintasun eta eskubide berberak dituztela.*

   a) Basque

   b) Breton

   c) Estonian

   d) Turkish

5. *Bütün insanlar hür, haysiyet ve haklar bakımındaneşit doğarlar.*

   (Unique letter: ı)

   a) German

   b) Maltese

   c) Albanian

   d) Turkish

6. *Kaikki ihmiset syntyvät vapaina ja tasavertaisina arvoltaan ja oikeuksiltaan.*

   a) Albanian

   b) Finnish

   c) Hungarian

   d) Norwegian

7. *Sva ljudska bića rađaju se slobodna i jednaka u dostojanstvu i pravima.*

   (Unique letter: đ)

   a) Albanian

   b) Icelandic

   c) Polish

   d) Croatian

8. *Il-bnedmin kollha jitwieldu ħielsa u ugwali fid-dinjità u d-drittijiet.*

   (Unique letter: ħ)

   a) Lithuanian

   b) Maltese

   c) Polish

   d) Turkish

9. *Pub den oll yw genys rydh hag kehaval yn dynita ha gwiryow.*

   a) Albanian
   b) Cornish
   c) Croatian
   d) Estonian

10. *Toate fiinţele umane se nasc libere şi egale în demnitate şi în drepturi.*

    (Unique letter: ţ)

    a) Catalan
    b) Sardinian
    c) Romanian
    d) Sicilian

11. *Kõik inimesed sünnivad vabadena ja võrdsetena oma väärikuselt ja õigustelt.*

    a) Estonian
    b) Finnish
    c) Sami
    d) Swedish

12. *Visi žmonės gimsta laisvi ir lygūs savo orumu ir teisėmis.*

    (Unique letter: ė)

    a) Croatian
    b) Maltese
    c) Lithuanian
    d) Turkish

8

Games of Tongues

INVENTED LANGUAGES

Languages are full of obscure rules, ambiguities, and irregularities. They can be difficult to learn, hard to pronounce, and confusing to write. For hundreds of years, lexical experimenters have tried to see if they could outdo the languages occurring naturally in the world by coming up with their own.

The first person documented to have invented a language was the twelfth-century German nun Hildegard von Bingen, one of the most prominent women in the medieval church. She drew up a vocabulary list of about a thousand words—from *Aigonz*, "God," to *zirzer*, "anus"—although, sadly, only two sentences of written text survive. In the centuries since, however, hundreds—if not thousands—more languages have been concocted, with their own extensive vocabularies, grammars, and sometimes even alphabets. We will encounter a few particularly interesting examples in this chapter. Let us marvel at the creativity and dedication of language constructors, and acquire a taste for the ingeniousness and beauty of their inventions.

Language creation seems to fulfill some kind of human urge. Since forever, for example, children have played language games. One of these games, Pig Latin, was popularized in the 1930s by the Three Stooges, and as a result some Pig Latin words have entered the American English dictionary. The title of the next puzzle, for example, is a line from the movie *The Lion King*.

71

IXNAY ON THE UPIDSTAY

Written below, in Pig Latin, are the names of five children's books:

> Arryhay Otterpay Andway Ethay Amberchay Ofway
>   Ecretssay
> Atildamay
> Ethay Uffalogray
> Ethay Ionlay, Ethay Itchway, Andway Ethay Ardrobeway
> Oodnightgay Istermay Omtay

What are the books' titles in English? And what is the name of this book in Pig Latin?

Games like Pig Latin come from fooling around, from children being children. Our main concern here, however, is languages that are designed by individuals with a clear purpose in mind. Such as Solresol, invented in the 1830s by the Frenchman Jean-François Sudre.

Sudre hoped that Solresol could become a universal language, spoken by everyone, no matter where they came from. Or indeed sung by everyone. Or whistled, or even played on a violin. Solresol, you see, is a language made up entirely of *do, re, mi, fa, sol, la,* and *si,* the syllables for the seven musical notes of a major scale. Sudre organized the vocab based on logical principles. The most common words are single notes: *do* means "no," *re* means "and," and—rather conveniently for Spanish or Italian speakers—*si* means "yes." Other common words are made up of two and three notes. All four-note words, however, are organized by category based on the initial note: the "key" of *do,* for example, is "dedicated to physical and moral man, his intellectual faculties and his qualities, and food." Thus *doremido*, "forehead"; *doremire*, "eyes"; *doremifa*, "nose"; and so on.

Sudre capped word length at five notes, and banned words with repetitions (like *dododo*). He calculated that his vocabulary, once completed, would have only 11,732 words, which is fine for daily use. He toured the concert halls of Europe giving demonstrations of Solresol, speaking it with his mouth and playing it on instruments. Such was his celebrity that, when he was in London, he met King William IV.

72

NAME THAT TUNE

Here are some phrases in Solresol and their translations in English:

Redo faresimi soldorela solsido fasimire	My black cat runs fast
La solmisire soldosoldo remisifa remi refaredo	The teacher slowly opens your (singular) wardrobe
Refa solmisire solfamido laredola la faresimi lafamido	His old teacher buys a little cat
La resolsoldo ladorela resol refasire laredosol	The carpenter sells our white box
La dofarela silami la sisifado	The rich man hates the lawyer

a) **Translate into English:**

   Resi sisifado ladorela la refaredo domifala

b) **Translate into Solresol:**

   (i) The young carpenter loves your (plural) cat
   (ii) The poor man quickly closes the black box

Europe in the late nineteenth and early twentieth centuries was a hotbed of language invention. As travel and trade between nations grew, thanks to new technologies like trains and factories, so did ethnic tensions. Among linguists, the idea took off that Europe needed a language that everyone could speak, something that would unite the continent by reducing the fear and misunderstanding between different groups. People began creating languages—Universalglot, Volapük, Weltsprache, and many more—that they hoped would fulfill this role. None of them achieved their goal of becoming an international auxiliary language. One of them, however, did achieve something remarkable: It became the only constructed language ever to have its own "native" speakers.

Esperanto was invented by Ludwik Zamenhof, born in 1859 in Bialystock, a city in present-day Poland. He grew up saddened that his German, Polish, Russian, and Yiddish neighbors never spoke to each other, their mutual antagonism a microcosm of Europe's as a whole. He started to devise Esperanto—it means "one who hopes"—while still in his teens. Zamenhof created a large and passionate community of speakers across Europe, and the language took on a life of its own. Esperanto currently has about 100,000 active speakers, of which 10,000 speak it fluently and about 1,000 have it as their mother tongue. Zamenhof's original mission of comradeship is still central to Esperanto identity. Learn the language and you get access to *Pasporta Servo* (Passport Service), a social network that provides free lodging for Esperanto speakers around the world.

Zamenhof designed Esperanto to be easy to learn, easy to pronounce and (in nineteenth-century terms) politically neutral. He adapted his core vocabulary from words in the major European languages. For example:

arbo	tree
birdo	bird
hundo	dog
leono	lion
libro	book
pomo	apple
rozo	rose
suno	sun

Grammar in Esperanto is simple, regular, and logical. Unlike every other natural language in the world, its rules allow no exceptions.

For example, every noun ends in an *-o*, every adjective ends in an *-a*, and every infinitive verb ends in an *-i*. All plurals end in a *-j*. There are no genders. Words are created by agglutination, that is, they are built up out of root words, prefixes, and suffixes.

To solve the following problem, you need to isolate the roots, prefixes, and suffixes in each of the given Esperanto words. Once you've done that you should be able to construct some Esperanto words of your own. One hopes.

73

A FIX OF AFFIXES

Here are some Esperanto words, together with their English translations:

aliam	at another time
aliel	otherwise, in another way
ĉiamo	eternity
ĉie	everywhere
kia	what kind (of)?
kie	where?
kielo	way, mode of action
kiomo	quantity
neniala	with no cause (causeless)
neniam	never
tial	therefore (that's the reason why)
tiama	of that time
tiom	that many

Translate into English:

1. alial

2. nenio

3. ĉiel

4. tie

Translate into Esperanto:

5. different

6. nowhere

7. omnipresent

8. like that (in that way)

Esperanto *is* easy to learn. A study of French secondary school students showed that after 150 hours of Esperanto they achieved the same standard of fluency that they got after 1,000 hours studying another Romance language, 1,500 hours studying English, or 2,000 hours studying German. Because of this, many schools around the world teach Esperanto as an introduction to learning foreign languages.

Yet the grammatical simplicity of Esperanto does not create poverty of expression. Alexey Pegushev, the Latvian linguist who devised the previous (and the next) problem, told me that, counterintuitively, the "treacherous simplicity of the grammar rules allows highly complex constructions. I would often make up words I had never used before by stacking suffixes and prefixes in a logical manner, and my collocutor would understand, and vice versa." Although, he admits, not everyone likes to push the envelope the way he does. "Many Esperanto speakers opt for maximum simplicity in order to stay true to the original goals of mutual comprehension and international brotherhood." Amen.

The next problem shows just how concise and efficient Esperanto is when it comes to verbs.

74

THE HUNGRY GOAT IS TENSE

Here are some sentences in Esperanto, and their English translations. The ĝ is pronounced like "g" in the word "gem."

1. *La kapro manĝintas*   The goat has eaten
2. *La kapro manĝitos*   The goat will have been eaten
3. *La kapro manĝis*   The goat ate
4. *La kapro manĝas*   The goat eats
5. *La kapro manĝintis*   The goat had eaten
6. *La kapro manĝotas*   The goat is going to be eaten
7. *La kapro manĝantas*   The goat is eating
8. *La kapro manĝontis*   The goat was going to eat
9. *La kapro manĝintos*   The goat will have eaten

Translate the following into English:

a) *La kapro manĝontos*
b) *La kapro manĝitas*

Translate the following into Esperanto:

c) The goat was eating
d) The goat is being eaten

The great seventeenth-century thinker Gottfried Leibniz wrote that if there was a universal symbolic language to express mathematical, scientific, and philosophical ideas, then human reasoning could be reduced to calculation. There would be no scientific disputes, since you would just plug ideas into the symbolic language and calculate the correct outcome. Historians of science say that Leibniz's idea foreshadowed much later concepts, such as symbolic logic and computer programming.

In 1921, the Estonian linguist Jacob Linzbach took Leibniz at his literal word, devising a way to write sentences using mathematical symbols and operations. He called his artificial language "transcendental algebra." I guarantee you have never seen equations as charming (or as utterly bonkers).

75

THE WICKED GIANT ATE THE PARENTS

Here are eight sentences in transcendental algebra, with their English translations:

1. $\left(\dfrac{\dot{\Lambda}\dot{\Delta}\dot{\iota}\dot{\Delta}}{\dot{\Delta}\dot{\iota}\dot{\Delta}} + \dfrac{\dot{\iota}\dot{\Delta}}{\dot{\Delta}} \right)^{\leqslant}$ The father and the brother are talking

2. $n(>\dot{\mathrm{I}})^{\square-}$ The giants are working

3. $\left(\dfrac{\dot{\iota}\dot{\Delta}(-\dot{\Lambda}\dot{\Delta})}{(-\dot{\Lambda}\dot{\Delta})} \right)^{\mathscr{O}} = \boxtimes$ The orphans are writing a letter

4. $(-n\dot{\mathrm{I}}_1)^{\mathscr{O}}$ It's not us writing

5. $\boxtimes^{\sqrt{\varnothing}} = -\dot{\triangle}_3$ It is not by her that the letter is being written

6. $\left(\dfrac{\dot{\wedge}\dot{\triangle}\iota\dot{\triangle}}{\dot{\triangle}\iota\dot{\triangle}}\right)^{-\heartsuit} = \boxempty\!-$ The father does not like the work

7. $(n(<\dot{I}) + \heartsuit)^{-\frown} = \dfrac{\dot{\wedge}\dot{\triangle}\iota\dot{\triangle}}{\dot{\wedge}\dot{\triangle}}$ The good dwarfs are not eating the children

8. $((>\dot{I}) - \heartsuit)^{\frown} - t = \dfrac{\dot{\wedge}\dot{\triangle}\iota\dot{\triangle}}{\iota\dot{\triangle}}$ The wicked giant ate the parents

Translate into English:

a) $(<\dot{I} + n(>\dot{I}))^{\overset{<}{}}$

b) $\left(\dfrac{\dot{\wedge}\dot{\triangle}\iota\dot{\triangle}}{\dot{\wedge}\dot{\triangle}\iota} - \overset{<}{}\right)^{\varnothing}$

c) $\dot{I}_3^{\heartsuit} - \sqrt{\heartsuit}$

d) $\boxtimes^{\sqrt{\frown}} - t = \dfrac{\iota\dot{\triangle}}{\iota} - \frown$

Charles Bliss invented a pictorial language when he was a refugee in Shanghai during the Second World War. Strolling the streets one day, he noticed that he interpreted the Chinese characters in shops and signs as words in German, his mother tongue. Following this realization, he created Blissymbolics, a visual language in which words are represented by simple symbols that can be read by the speakers of any language. He designed the symbols to resemble the things they were representing as closely as possible. So, for example:

mouth

o

nose

∠

Basic symbols combine to make more complicated words. In order to distinguish between parts of speech, a circumflex (^) indicates a verb, and an inverted circumflex an adjective.

Bliss's motivation, like that of so many other language inventors, was to secure world peace. Born in the Austro-Hungarian Empire, he was imprisoned in both Dachau and Buchenwald concentration camps at the beginning of the Second World War. Traumatized by the conflict, he wanted to design a utopian system that would bring people together and inhibit any future nationalist wars.

Bliss published an introduction to Blissymbolics in 1949, by which time he was living in Australia. It failed to garner any interest, even after his wife sent 6,000 letters to universities around the world. By 1970, Bliss was a widower in his seventies, resigned to the fact that he had failed to improve humanity, and that his life's work, ignored and unused, had been a waste of time.

Unbeknown to Bliss, however, his system *was* being used, and *was* transforming people's lives. A Canadian special-needs teacher had found a copy of his book and was using it with severely disabled children. For students who were unable to speak and had limited control of their hands, Blissymbolics was revolutionary. It allowed them to express themselves in a way they had never been able to before. The teacher showed the students a board displaying all the symbols, and by pointing at symbols in a sequence (or by indicating where the teacher should point), the students had the sense of speaking in a sentence for the first time.

Bliss was overjoyed when he received a letter from the teacher thanking him for his work. He flew to Canada to meet her and her students. Since then, the use of Blissymbolics by special-needs teachers has spread around the world. Margareta Jennische, a retired associate professor of speech-language pathology at Uppsala University in Sweden, is the president of Blissymbolics Communication International. She says that the system remains an invaluable aid to hundreds of people. Of the many picture-based systems that disabled people use to communicate, she says, only Blissymbolics is a language. "The symbols are not too abstract. But they enable you to express abstract concepts. It is an amazing tool and could be used so much more."

76

THE FACE OF BLISS

Shown below are 12 words written in Blissymbolics, and their English translations in random order. Remember, in Blissymbolics a circumflex (^) indicates a verb, and an inverted circumflex an adjective:

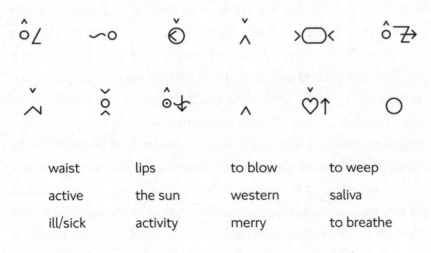

waist	lips	to blow	to weep
active	the sun	western	saliva
ill/sick	activity	merry	to breathe

a) Work out the correct correspondences.

b) Translate the following symbols. (Two have the same meaning.)

c) Write in Blissymbolics:

air   torso   to rise   east   sad

Yukio Ota is a Japanese graphic designer, now in his eighties, whose most famous work you are certainly familiar with. He designed the "running man" exit sign: the green character dashing through a white door, with one leg cocked back. It's the global standard for exit signs.

Yet if you ask Ota what work he is most proud of, he will say his pictorial language, the Lovers' Communication System, or LoCoS.

In the 1960s and '70s, Ota corresponded with Charles Bliss about their shared idealism for universal languages. Yet Ota felt that Blissymbolics had a serious shortcoming. It was too European, because it reads from left to right, along a single horizontal line, with spaces between the words, and thus "clings to the shores of the culture from which it came."

LoCoS arranges its symbols in a more Asian way. There are no spaces between symbols, just as there are no spaces between Chinese and Japanese characters. And just as Chinese characters can be modified by adding elements in two dimensions, so LoCoS expands up and down, as well as across. "Readers can make up their mind about the best way to perceive the meaning," Ota says, "just as they do when looking at a frame in a movie."

Ota told me of some Russian undergraduates who did an assignment on LoCoS in 2019, and a Japanese science museum that recently ran a course. These small and sporadic bursts of interest fill him with enthusiasm that LoCoS may one day be used across the world.

77

BIG FISH LITTLE FISH

Here are some simple sentences in LoCoS, translated:

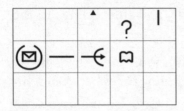

Tomorrow the fisherman will see many fish.

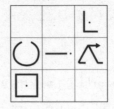

This man will go later.

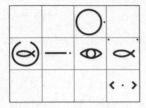

The postman is giving a book to Carina.

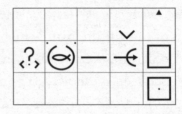

How many fisherman give Vika this thing?

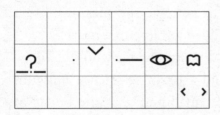

Where did Eva see the big book?

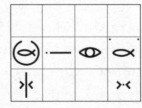

The thin fisherman saw few fish.

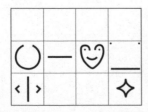

The fat man likes pretty places.

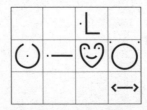

Beforehand I used to like long days.

a) Translate the following sentences:

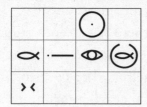

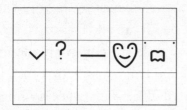

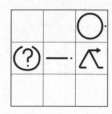

b) Write the following sentences in LoCoS in the grids given below:

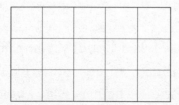

Yesterday Vika saw a small fisherman.

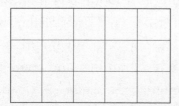

What will the reader give me?

The creators of the languages we have considered so far all wanted their inventions to be used by as many people as possible.

In the last few decades, however, there has been a surge in creators who have no interest in anyone actually *learning* the languages they are painstakingly constructing. The point of these linguistic endeavours is artistic, to enhance a fictional universe.

J. R. R. Tolkien was the first major novelist to show just how powerfully attending to linguistic detail can add realism to an imagined world. In fact, Tolkien always said the point of his fictional universes was to enhance the fictional languages, rather than vice versa. Perhaps the most successful of all art-languages is Klingon, the mother tongue of *Star Trek*'s belligerent, crinkly-foreheaded alien species. Its inventor, Marc Okrand, who has a PhD in linguistics, wanted it to sound as unlike a human language as possible. So he mixed sounds that don't usually go together, and gave the grammar quirks like an object-verb-subject word order. Yet Klingon (or *tlhIngan Hol*, as it is called by native speakers) has been an unexpected success in the real world. Klingon has its own language institute, which runs conferences and promotes the language in social media and through language-learning software.

In 2009, David J. Peterson was asked by a TV company to construct a language for the Dothraki, an uncouth yet honorable tribe of scantily clad, horse-riding fighters who would be featuring in a forthcoming series, *Game of Thrones*. What sort of guttural sounds might come out of a Dothraki's mouth? Peterson decided on a phonology of trilled "r"s, back-of-the-throat "q"s, tongue-on-the-front-teeth "t"s, and certainly no prissy "p"s or babyish "b"s. He also fleshed out a grammar and etymology. Dothraki sounds like a proper language, because—on paper— it is.

78

BLOOD OF MY BLOOD

Here are nine words in Dothraki and their English translations, both listed in alphabetical order (which means that the Dothraki words are probably not alongside their English translations):

ajjalan	autumn
chafka	blood of my blood
jalanqoyi	dragon
qoy qoyi	funeral pyre
shekh	lunar eclipse
shekhqoyi	solar eclipse
vorsaska	summer
vorsqoyi	sun
zhavorsa	tonight

What are the Dothraki words for "blood," "fire," "lizard," "wind," and "moon"?

a) qoy, jalan, chaf, vorsa, zhav

b) qoy, zhav, jalan, chaf, vorsa

c) zhav, qoy, chaf, vorsa, jalan

d) vorsa, chaf, jalan, qoy, zhav

e) qoy, vorsa, zhav, chaf, jalan

To solve this problem, it's helpful to draw a map of the Dothraki words, as shown on the next page. I've highlighted the five root words in boxes. The translation of one of the root words is fairly clear, and that should give you a foothold. To complete the puzzle, look at how the root words combine in the longer words, and then compare this information with how the English meanings share common ideas.

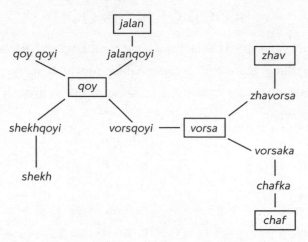

The Klingon and Dothraki languages were both invented by men, for the purposes of being spoken by hypermacho warrior races.

Suzette Haden Elgin, on the other hand, devised Láadan for women. Elgin was a science fiction writer and a linguistics professor. In the early 1980s, she decided to explore the hypothesis that existing languages are intrinsically misogynistic. She wrote a novel, *Native Tongue*, about a group of women who create a language, Láadan, for the purposes of expressing the perceptions of women. Genders are reset to make the feminine the default: The word for "woman" is the basic form of the word for "person," and the word for "man" is the word for "woman" plus a suffix to denote the male form. Láadan has an expanded vocabulary for female physical experiences, including six different words to describe the experience of menstruation (*ásháana* = "to menstruate joyfully"). It also has a vocabulary for emotional context, such as words that indicate on what grounds information is known (*waá* = "I assume it's false because I don't trust the source").

For Elgin, Láadan was as much a scientific experiment as a piece of science fiction. She wanted to illustrate the hypothesis that the structure of a language influences the way its speakers view the world, and she also wanted to see whether women would welcome and nurture the idea of a women's language. *Native Tongue* became a cult book, but her language never captured the imagination of more than a few diehards. A few years before her death in 2015, she ruefully noted that Klingon, however, goes on and on.

79

WITH AND WITHOUT

Here are six words in Láadan and their English translations, both listed in alphabetical order:

ewith	anthropology
háasháal	crowd
háawith	morning
hanesháal	ream of paper
mémel	Thursday
méwith	young person

Match the words to their correct translations.

For this problem, draw a map of the English words linked by meaning. I've started it off for you below. Members of the group on the left are linked by the idea of people, and those on the right by the idea of time. When you work out how to join the two fragments, you'll see that a map of the Láadan words and the map of the English words coincide.

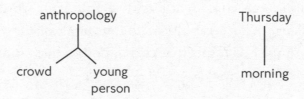

Toki Pona is the "cutest" of all invented languages. Or so I was told when I asked members of the 4,500-strong Toki Pona Facebook group why it has become such a darling within the constructed-language community. Toki Pona is a minimalist language, created by the Canadian linguist Sonja Lang in 2001. It has a vocabulary of only about 120 root words—a number small enough to be memorized in a weekend. For example, there are five words for colors: "black," "white," and the three primary colors, and other colors can be expressed by combining these terms. Toki Pona words derive mainly from English, Tok Pisin, Finnish, Esperanto, Croatian, Dutch, and a few other languages.

Toki Pona uses nine consonants—*j, k, l, m, n, p, s, t*, and *w*—and five vowels: *a, e, i, o*, and *u*. Double consonants are banned. When translated into Toki Pona, a proper name in English that uses one of the forbidden consonants would be rewritten with the most similar-sounding permitted letter, so, say, "v" becomes *w*, and "f" becomes *p*.

Lang did not purposefully set out to make her language cute; it was, rather, a philosophical experiment in reducing a language to its bare essentials. Speaking Toki Pona is like mental yoga, in that you try to express as much as you can using the smallest possible number of fundamental elements. Indeed, Lang invented the language as a way of simplifying her own thoughts, and inducing positive thinking, during a bout of depression. Toki Ponians say they enjoy its "rustic simplicity" and the way it forces you to think about meaning. And the fact that you can learn it fast, before any tedium sets in.

In the following problem, the Toki Pona translations of most of the English terms contain at least two Toki Pona words. Part of the challenge is to break down the meaning of the English words into simpler concepts that correspond to these individual words in Toki Pona.

80

PISS AND GOLD

Here are 12 terms in Toki Pona and their translations in English. Both lists are in alphabetical order.

ilo suno	*lipu toki*	boat	movement
jan ilo	*tawa*	book	piss
jan Powi	*tomo moku*	Boris	prophet
jan toki	*tomo tawa telo*	gold	restaurant
kiwen	*telo jelo*	ice	robot
kiwen suno jelo	*telo kiwen*	lantern	rock

Using the word map below, match the Toki Pona terms to their correct translations.

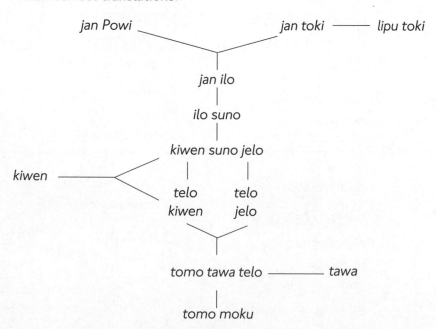

Our tour of linguistic innovation is not quite over. Prepare to meet the superstars of some super scripts.

LINGO BINGO

Pseudo Loanwords

Each of the words shown below is an English word that has entered a foreign language, where it has taken on a new meaning.

Can you guess the correct meaning of each of these "English" words in the languages given?

1. DRESSMAN (German)
    a) Drag queen
    b) Fashion designer
    c) Male model
    d) Tailor

2. ESKIMO (Italian)
    a) Parka
    b) Igloo
    c) Husky
    d) Snow shoe

3. VIKING (Japanese)
    a) Sailor
    b) Blonde
    c) Tourist
    d) All-you-can-eat buffet

4. SMOKING (French)
    a) Cigar
    b) Tuxedo
    c) Sexy person
    d) Pollution

5. MONKEY CLASS (Danish)
    a) Disruptive children
    b) Zoo cage
    c) Call center
    d) Economy class

6. OUTDOOR (Brazilian Portuguese)
    a) Patio
    b) Billboard
    c) Toilet
    d) Service exit

7. FOOTING (Spanish)
   a) Pedicure
   b) Shoeshop
   c) Kickboxing
   d) Jogging

8. ONE SHOT (Korean)
   a) Bottoms up!
   b) Revolver
   c) Penalty kick
   d) Vaccination

9. SLIPPERS (Dutch)
   a) Curling
   b) Flip-flops
   c) Underwear
   d) Eels

10. VAUXHALL (Russian)
   a) Car
   b) Railroad station
   c) Gay bar
   d) Hat

9
Script Tease

CONFESSIONS OF AN
ALPHABET ADDICT

Awriting system is a thing of beauty. I never appreciated quite how true this is before I began the research for this book, during which I cast my eyes over hundreds of scripts previously unknown to me. In this chapter I will feature ten of my favorites. Some made the cut for their grace and style, some for their ingenuity, and others for the incredible tales they tell.

We start with the remarkable story of the Native American who invented his own writing system without prior knowledge of reading. Sequoyah grew up in the late eighteenth century speaking only Cherokee, which, like all the indigenous languages of North America at that time, was only spoken, not written down. Sequoyah saw that the white settlers had a way to communicate their language using symbols on paper—what he called "talking leaves"—but since he could not read English, the meaning of these symbols was a mystery to him. Still, he appreciated that the written word was a source of power, and he set out to create a similar system for his own language.

How does one create a script from scratch? In his first attempt, Sequoyah had a different symbol for every word. Too cumbersome. His second attempt was a syllabary, with a symbol for every syllable. (The progression from word-symbols to syllable-symbols also took place in Sumerian cuneiform and Egyptian hieroglyphics.) Sequoyah had no education in linguistics, nor a writing system available to him, which made isolating the syllables in Cherokee a formidable task. He built a cabin for himself, in which he spent his days obsessively working on his project. According to one story, his wife burned the cabin down because he spent so much time there.

Sequoyah eventually listed 86 syllables and assigned them characters. His achievement made him the earliest known example of a person from a preliterate society to devise an original, fully functioning writing system. He borrowed a dozen Latin characters for his syllabary, as well as the number 4, from a copy of the Bible, although since he did not read English, none of his characters are pronounced the same way. In his day job, Sequoyah was a silversmith, which perhaps explains why the other characters he chose are highly ornate, with many decorative typographical features, such as beaks, balls, and serifs. In 1825, this syllabary became the official Cherokee script, and in 1827 the first font was created, and a printing press shipped to the Cherokee Nation in Oklahoma. Sequoyah's fame spread across the world. One botanist paid tribute by naming a genus of giant redwood trees—the "sequoias"—after him. (At least, this is the most likely story of how the trees got their name. The etymology has recently been disputed.)

Sequoyah's syllabary had great social and political impact. Cherokee speakers found it easier to learn than English speakers found the alphabet, because the syllabary was tailored to the Cherokee language and had none of English's peculiar pronunciation rules. As a result, the Cherokee became literate at a faster rate than their white neighbors, and the script was an important unifier for the tribe at a time when its members were dispersed across the country.

With about 300,000 members, the Cherokee tribe is the largest federally recognized Native American tribe in the US. Even though the Cherokee language is critically endangered, with only a couple thousand speakers left, the syllabary remains an important marker of Cherokee identity and has enjoyed a resurgence in the last few decades, helped by the politics of indigenous self-governance, as well as by technology. Despite its relatively few speakers, it is in common use on street signs, in books, and online.

81

CORNSILK AND THE DEVIL

Here are 12 words in Cherokee and their pronunciations, arranged in random order. Match the words to their correct pronunciations. (The translations are provided for information only, and do not contribute to the solution.)

1. ᏍᏅᎯᎣᏒ	a)	*tsvsgina* (wind)	
2. ᎤᏌᏟ	b)	*etsi* (egg)	
3. ᏝᎷ ᎤᏞᎥᏗ	c)	*atsilvsgi* (cornsilk)	
4. ᎠᏩ	d)	*gola* (fire)	
5. ᎠᏟᎣᎩ	e)	*uwetsi* (flower)	
6. ᎠᏎᏯ	f)	*tsatiyosidi* (money)	
7. ᎡᎯᏍᎥᏗ	g)	*uyona* (approve)	
8. ᎤᏏᎾ	h)	*selu unenudi* (horn)	
9. ᏎᏟ	i)	*atsila* (mother)	
10. ᏣᎣᏯᎾ	j)	*ganolvvsgv* (devil)	
11. ᎠᏩ	k)	*adela* (bone)	
12. ᏣᏟᏏᏗ	l)	*gohiyudodi* (quarrelsome)	

Fifteen hundred miles due north of the Cherokee Nation, on the northern banks of Lake Winnipeg in Canada, is the small town of Norway House, formerly a fur-trading post for the Indigenous Cree people (see map on page 175).

In 1840, James Evans, a Methodist minister from Hull in England, arrived in Norway House to work as a missionary. Inspired by Sequoyah's Cherokee syllabary, Evans set about devising one with the Cree, who also had no system for writing down their language. Unlike Sequoyah, however, Evans was a knowledgeable amateur linguist, and he borrowed ideas from other writing systems, such as the Indian Devanagari alphabet, which provided the shapes of some of his characters. For example, Devanagari ट (for *ta*) became C. Evans also copied innovations from Sir Isaac Pitman's recently published shorthand. These included simple geometrical strokes, varied in thickness to indicate voiced and unvoiced consonants (although this element was soon lost due to the difficulties of printing hard and soft lines) and a curious feature that you will discover in the next problem.

Evans's syllabary was straightforward enough that Cree speakers were able to master it in only a few days, and it could be taught easily by family or friends. Within two years, the Cree leaders were using it, and within a decade it had spread to the Atlantic coast 2,000 miles away. By the late nineteenth century it was said that the Cree were among the most literate peoples in the world. The speakers of dozens of other Indigenous Canadian languages—spoken from coast to coast—adapted the syllabary for their own languages and dialects, and by the mid-twentieth century "Canadian Aboriginal Syllabics" had become an important part of Indigenous identity and heritage. Nowadays, even in those groups where Syllabics has fallen into disuse for daily communication, it is still used for ceremonial purposes, such as on artwork, plaques, and public monuments.

In one frozen outpost, however, Syllabics is thriving. Nunavut is the largest and most northerly Canadian territory, with a population of about 35,000. The territory has two official scripts: the Latin alphabet, used for English and French, and Inuktitut Syllabics, used for Inuktitut, the predominant Inuit language spoken there. The legislature publishes its proceedings in English and Inuktitut (in Syllabics). Inuktitut Syllabics is taught in schools as a subject in its own right and as a medium for teaching other subjects. Almost two centuries after Evans devised his Hindi-inspired shorthand, it survives as a living script in the polar north.

82

A COOL CALENDAR

Here are the months of the year in Inuktitut, written in Syllabics, together with their pronunciations. Add the three missing months in Inuktitut Syllabics.

Syllabics	Pronunciation
ᔭᓄᐊᕆ	jaannuari
ᕕᕝᕗᐊᕆ	viivvuari
ᒫᑦᓯ	maatsi
ᐃᐳᕆ	iipuri
	mai
ᔫᓂ	juuni
ᔪᓚᐃ	julai
ᐊᒡᒌᓯ	aaggiisi
ᓯᑎᐱᕆ	sitipiri
	utupiri
ᓄᕕᐱᕆ	nuvipiri
	tisipiri

Between the beginning of the nineteenth century and the middle of the twentieth, West Africa was a global center of typographical innovation. Its peoples, who spoke languages with no writing systems of their own, came up with dozens of indigenous scripts as a riposte to alphabetical colonization by Muslims and Christians using Arabic and Latin letters. Only two scripts from this era remain in use. One is a syllabary for the Vai (or ꕙꔤ) language of Liberia, which dates from the 1830s and was the first of West Africa's invented writing systems. Some scholars believe it came about because a Cherokee who immigrated to Liberia told the people there about Sequoyah's syllabary. Another origin story is that the idea came to a member of the Vai people in a dream. Whichever explanation is true, the Vai syllabary is on life support, and were it not for academic interest, it might have died out altogether.

The other West African script that survives is the N'ko alphabet, which is used for writing Manding, a group of languages spoken by about 30 million people in Guinea, Burkina Faso, Mali, and the Ivory Coast. It was created by the Guinean thinker Solomana Kanté in 1949, a time when West Africa was gaining independence from France. Kanté had two motivations: buoying African pride, and translating the Koran. Once he completed N'ko—which means "I speak" in Manding—he immediately put the alphabet to work, publishing about 100 books in it, including a 30,000-entry Manding-Manding dictionary and a history of West Africa.

N'ko has been a great success. More books are now printed in N'ko than in the official Latin orthography for Manding. The alphabet is used by tens of thousands of people, and you can see it increasingly in shops, on signs, and on the internet. French is the language of government in Guinea, Ivory Coast, Burkina Faso, and Mali, but most Manding speakers don't speak French, nor use the Latin alphabet. If a Manding speaker wants to learn how to read, N'ko is the way to n'go.

83

DESTINATION TIMBUKTU

Below are 12 West African place names written in the N'ko alphabet, along with their transliterations in arbitrary order. (The names in parentheses are the English spellings of these cities.) Note that N'ko is a cursive script in which the letters are joined by a horizontal line. Match the names in N'ko to the correct transliterations.

In the N'ko alphabet, the word "N'ko" is ꕢꕋꕫ, and "Kanté" is ꕐꔦꕫꕋ.

1. ꔤꕸꔰꕢꕐ   a) *kɔnakri* (Conakry)

2. ꕫꗞꕸꔰꕋ   b) *kindiya*

3. ꔱꕫꕚꔤꕸ   c) *n'srégbdɛ*

4. ꔤꕶꕐꕸꗞ   d) *sromaya*

5. ꖃꕚꖉꕫꕶꖹ   e) *faranna*

6. ꔤꕔꖃꔱ   f) *jikuɛ*

7. ꕫꗞꕸꕚꕋꕶꖹ   g) *tonbtu* (Timbuktu)

8. ꔤꖉꗞꖃꕋ   h) *bisawo*

9. ꖚꗞꕸꕋꖝ   i) *abijan* (Abidjan)

10. **ⵓⴱⴼⴶⴱ**   j) *npraeso*

11. **ⵠⵙⵣⵠⵙⵜⵐⵕ**   k) *gɛsoba*

12. **ⵐⵍⵕ̇ⵣⵝ**   l) *gkedu*

**In the transliterations, the letters ɔ and ɛ are vowels. I've
also left out the tonal diacritics usually present in N'ko.**

The Akan people of West Africa wear their hearts on their sleeves.
To be more precise, they wear clothes printed with striking,
highly stylized symbols that reflect how they see the world. These
symbols—called *adinkra*—each represent a saying, a proverb, a
moral value, or a philosophical concept, which makes them a
writing system of sorts. Here are two examples:

The Akan people live in Ivory Coast, Ghana, and Togo, and
speak Twi. In the illustration above, the *adinkra* on the left is
the *Asase ye duru*, which is Twi for "the Earth has weight." One
heart supports another, representing the importance of Earth in
sustaining life. The *adinkra* on the right is the *Woforo dua pa
a*, which is derived from from the saying *Woforo dua pa a, na
yepia wo*, which means "When you climb a good tree, you are
given a push." In other words, when you do good work, you will
get support. The symbol—which could be someone with their
arms around a tree trunk—represents support, cooperation, and
encouragement.

Traditionally, *adinkra* were only printed on robes worn at funerals. In the last few decades, however, they have become popular as prints on everyday clothing, pottery, and furniture. There are about 500 documented symbols, and in the next question we will encounter eight more of them, each of which contains visual clues as to what it might mean.

84

TWISTING TWI TWISTERS

Match the *adinkra* symbols shown below to their correct names and descriptions. Given that *ne* means "and," what are the probable meanings of *osram* and *nsoromma*?

a)

b)

c)

d)

e)

f)

g)

h)

1. *Ananse ntontan*: the spider's web. The spider (*ananse*) is the most common insect in African folktales. It is small, but thanks to its ingenuity it can trap larger prey. The symbol represents wisdom, craftiness, creativity, and the complexities of life.

2. *Nyansapo*: the knot of wisdom. The Akan say, "a knot made with wisdom can only be undone by the wise and not the foolish." The knot is thus a symbol of wisdom, ingenuity, intelligence, patience, and the power and skill of oratory.

3. *Sankofa*: return and get it. This symbol represents the importance of learning from the past.

4. *Funtunfunefu denkyemfunefu*. This image is of two animals fighting over food, even though they share one stomach. The symbol is a warning that infighting and tribalism harm all who engage in them. In other words, "people who share the same destiny should not fight."

5. *Nkyinkyim*: twistings. This symbol represents one's ability to adapt and change to the twists and turns of life. It's about resilience, initiative, dynamism, and versatility.

6. *Duafe*: the name of a tool. This is sometimes used to symbolize abstract qualities linked to femininity, such as love and care. It is also sometimes used to symbolize more concrete qualities, such as looking one's best and maintaining good hygiene.

7. *Akofena*. This image symbolizes courage, gallantry, valor, and heroism. It is also a symbol of state authority and the rule of law, and was a popular motif in the heraldic shields of many former Akan states.

8. *Osram ne nsoromma*. This symbol reflects the harmony, love, and interdependence between a husband and wife.

The Caucasus mountain range between the Black and Caspian Seas (see map on page 166) is a region of unusually rich linguistic diversity, so much so that medieval Arab geographers described the area as the "mountain of tongues." It is home to about 50 native languages, from seven different language families. It is also home to two beautiful alphabets, Georgian and Armenian, which are used nowhere else in the world.

Georgia and Armenia are neighbors. Their national languages are not linguistically related. Yet they both have their own alphabets, a claim that very few other countries in the surrounding areas of Europe and the Middle East can make. Indeed, the two scripts are a curious pair. Georgian letters are chubby and round, with extravagant swashes and dramatic diagonal tails. Armenian letters are upright and serious, with understated vertical ascenders and descenders. It's almost as if the scripts are in a dialogue with each other, stubbornly trying to be as different from the other as possible.

Both alphabets emerged just over 1,500 years ago. Mesrop Mashtots, an Armenian theologian, devised the Armenian alphabet in 405 CE. He wanted a script that properly captured the sounds of Armenian, so a local version of the Bible could be written. The Georgian alphabet dates from the same time, although no one knows who the inventor was, or indeed if it was a single person.

The alphabets are important features of Georgian and Armenian identities. Drive for half an hour outside the Armenian capital, Yerevan, for example, and you'll find the Armenian Alphabet Monument, a sculpture park consisting of the 39 letters of the Armenian alphabet carved from stone. If it wasn't on your bucket list, it is now.

85

GEORGIA ON MY MIND

Here are the names of three countries written in Georgian, with their English translations:

ბრაზილია — Brazil

პერუ — Peru

ურუგვაი — Uruguay

What are the names, in English, of these two countries?

1. არგენტინა   2. კოლუმბია

86

A YEAR IN YEREVAN

Here are the months of the year in Armenian. They sound similar to how they are pronounced in English. Which month is which?

օգոստոս	մայիս	հոկտեմբեր
ապրիլ	հունվար	սեպտեմբեր
դեկտեմբեր	փետրվար	հունիս
մարտ	նոյեմբեր	հուլիս

The alphabet as a concept was invented from scratch only once, in ancient Egypt, as we saw in chapter 4. Alphabets spread northward into Europe and eastward across Asia, eventually reaching the islands of Sulawesi, Java, and Bali in Indonesia, which were the farthest-flung places to have them before the Age of Discovery.

The alphabets of south Indonesia are among the most beautiful I discovered while researching this book—especially Javanese, which is the subject of the next puzzle. Javanese follows the same alphasyllabic system as Hindi's Devanagari (to which it is distantly related), which we also covered in chapter 3 (page 68). To recap: Consonants are represented by letters, and vowels by diacritics. Each letter-and-diacritic combination constitutes a unit that represents a syllable, with the consonant followed by the vowel. If there is no diacritic, the consonant is followed by the vowel sound "a." What makes Javanese such an attractive script is the glorious extravagance of its diacritics: As you are about to find out, it is hard to distinguish them from the letters.

Around 80 million people speak Javanese. Sadly, however, in the last few decades the Javanese alphabet has been almost entirely replaced by an official Latin orthography.

SORCERY IN SOUTHEAST ASIA

Opposite are ten words in Javanese and their pronunciations, arranged in random order. Match the words to their correct pronunciations:

1. ᬨᬬᬃ　　　　　　a) *ngajar* (to teach)

2. ᬨᬶᬚᬸᬮ　　　　　b) *borang* (trap using bamboo spikes)

3. ᬢᬼ　　　　　　　c) *teluh* (sorcery)

4. ᬤᬵᬤ　　　　　　d) *kesiring* (shaved)

5. ᬚᬸᬭᬸᬄ　　　　　e) *juruh* (molasses)

6. ᬩᬩᬤ　　　　　　f) *babad* (history)

7. ᬧᬾᬦ᭄ᬢᬶᬮ᭄　　　g) *malikat* (angel)

8. ᬚᬯ　　　　　　　h) *jawa* (Java)

9. ᬬᬸᬫᬦᬶ　　　　　i) *pentil* (fruit bud)

10. ᬓᬲᬶᬭᬶᬂ　　　　j) *yumani* (underworld)

Users of the alphabets of southern Indonesia originally wrote the letters on a dried palm leaf with a knife-stylus, before rubbing the leaf with a material whose ink soaked into the grooves. In Java, these leaves were collated, perforated, and threaded together with string to create "manuscripts."

In southern Sulawesi, on the other hand, palm-leaf manuscripts looked much more modern: The leaves were cut into long, thin strips, stuck end-to-end, and then wound around a wheel like a cassette tape.

The language of southern Sulawesi is called Bugis, and its alphabet is called Lontara script. Even though Lontara is derived from the same predecessor as the Javanese alphabet, and both scripts are alphasyllabaries, they could not be more different. Javanese is majestic and flowing, decorated with copious diacritics, and provides everything you need to know to be able to pronounce it properly. Lontara, on the other hand, is stark and minimalist, giving you almost no help at all. The chevron-style consonants look almost identical; there are no spaces between words; and it uses only one punctuation mark. In addition, if a syllable has a final consonant, the character for this consonant sound is omitted. Interpreting Lontara is a puzzle even when you know how to read it. (Be warned. The next puzzle is a toughie.)

One of the many reasons Lontara script is of interest to scholars is that it was used to transcribe the longest written work in world literature. The *Sureq Galigo* is an epic poem written in pentameter, and details the creation myth of the Bugis people. The poem was passed down as an oral tradition until a century ago, when a Dutch missionary asked some local scribes to write it down. When they did, it ran to 7,000 pages. (Part of it is now kept in the Leiden University Library.)

The *Sureq Galigo* begins with the Lord of the Upper World looking for his servants. When he finds them, he discovers that they have visited the Middle World, aka Earth, which, to their surprise, was uninhabited. As a result, the Lord sends his son to become the first human on Earth. (With presumably hilarious consequences.)

You can still see Lontara in south Sulawesi. Street signs are bilingual Indonesian/Bugis, the former written in the Latin alphabet and the latter in Lontara. The script is also still taught in schools.

A Lontara "manuscript" from around 1900 in which the palm leaves are scrolled like a cassette tape. (Picture: Sirtjo Koolhof.)

88

BLAME IT ON THE BUGIS

The following text in Lontara script is from the opening of the *Sureq Galigo*. (The symbol ⸪ is a punctuation mark used for both commas and periods.)

Here is the same passage, translated into English:

> There is no one to call the gods, Lord, or to offer praise
> to the underworld. Why, Lord, don't you have one of
> your children descend, and incarnate him on the earth;
> do not leave the world empty and the earth uninhabited.
> You are not a god, Lord, if there are no humans under the
> heavens, above the underworld, to call the gods, Lord.

In the text below, the Bugis is written in transliteration, in lines of either 10 or 15 syllables. The lines are listed in random order.

A. *ajaq naonro lobbang linoé*

B. *lé namasuaq mua na sia*

C. *makkatajangeng ri atawareng.*

D. *mappaleq wali ri pérétiwi.*

E. *mattampa puang lé ri batara,*

F. *mattampa puang lé ri batara.*

G. *ri awa langiq, lé ri ménéqna pérétiwié,*

H. *tabareq-bareq ri atawareng,*

I. *tammaga puang muloq séuwa rijajiammu,*

J. *teddéwata iq, puang, rékkua masuaq tau*

Arrange the lines in their correct order.

How do you find a word in a Chinese dictionary?

Chinese writing is a system that uses thousands of different characters, with no alphabet underpinning how the characters are written. There is no way of being certain how any one individual character would be pronounced if you haven't seen it before (although a familiar speaker can have a good guess).

Instead, Chinese dictionaries categorize the individual characters that make up words into about 200 groups of "radicals," which are the basic components of any one character. For each radical, the characters are listed by the number of "strokes" it takes to write the rest of the character (having already taken care of the radical component).

The stroke is the basic element of Chinese writing. All characters have a stroke order. Indeed, when a child in China is learning to write, the stroke order is what is important, since it provides the calligraphic template for the character. For example, stroke order enables hastily written characters to be read, since the reader will decipher the scribble by imagining the order in which the strokes were made.

Stroke order is determined by a few principles, some of which you are about to deduce.

89

THE STROKES

The Chinese characters for the numbers 1 to 10 are:

一　二　三　四　五　六　七　八　九　十

These characters are written using this stroke order:

1. 一

2. 二

3. 三

4. 四

5. 五

6. 六

7. 七

8. 八

9. 九

10. 十

Assuming that stroke order follows a logical system, what are the correct stroke orders for the following characters?

木
mu, wood
(4 strokes)

白
bai, white
(5 strokes)

面条
miantiao, noodle
(16 strokes)

Chinese is one of the world's hardest writing systems to learn, since it requires you to memorize thousands of characters. The Korean alphabet, or Hangul, is one of the easiest. It is widely lauded by linguists as the most perfect, rational, and scientific script ever devised.

Hangul was invented in the 1440s by a team of Korean scholars—the Hall of Worthies—under the direction of King Sejong, a leader who invested widely in science, technology, and the arts. At that time, Korean used Chinese characters, which took a long time to master. National literacy levels were low. Sejong wanted a system for the Korean language that could be picked up by anyone relatively quickly. The result, published by royal decree in 1446, was Hangul, of which one of the Worthies wrote: "an intelligent man can acquaint himself with [the letters] before the morning is over, and even a simple man can learn them in the space of ten days."

Hangul is an alphabet with a number of linguistic enhancements. Words are presented in blocks of single syllables, so you can instantly break down the pronunciation. (With linear alphabets like Latin, by contrast, it's not always clear where the syllables fall in particular words.) By looking at the shape of a symbol, you can tell whether it is a consonant or a vowel. Consonants are, in fact, designed to resemble the way the sound is produced in the mouth. For example, the symbol ㄱ means "k" because your tongue makes the shape ㄱ when you pronounce it (imagine a sideways cross section of your tongue when you are facing left, where the back of the tongue touches the roof of the mouth). Similar sounds have similar symbols, so the "kh," which is a "k" spoken with an extra burst of air in the throat, is written with an extra line: ㅋ.

King Sejong introduced Hangul as an alternative to Chinese characters for the education of the masses. For a while, however, the Korean elite were contemptuous of what they saw as a common script for common people. Yet since Hangul was so easy, it took root, and five centuries later, in 1894, it became the country's officially sanctioned writing system. Now Koreans are so proud of their alphabet that they have a public holiday in its honor. Hangul Day is October 9, and commemorates the date in 1446 when Sejong first presented his creation.

GANGNAM STYLE

Below are four well-known place names in Korea, written in Hangul and English. The Korean names are not listed in the same order as the English ones.

1. 서울   a) Busan

2. 부산   b) Gangnam

3. 평양   c) Pyongyang

4. 강남   d) Seoul

Match the English and Korean names, then work out what this says:

5. 삼성

If you get this correct, you may find it fitting to celebrate your achievement with a rendition of Schubert's *Die Forelle*.

LINGO ⓑ ⓘ ⓝ ⓖ ⓞ

Untranslatables

The words and phrases in the following problems are sometimes considered "untranslatable," since there are no precise words or phrases in English that convey the exact meaning of the original language. Although, of course, I have translated them for you here.

Can you work out the correct meanings from their literal translations?

1. KUMMERSPECK (German)
   Literal translation: *grief bacon*

   a) The fat you put on from overeating when you are sad
   b) Comfort food served at funerals
   c) Piece of meat eaten by a vegetarian
   d) The longing for meat held by recent converts to vegetarianism

2. ESPRIT DE L'ESCALIER (French)
   Literal translation: *wit of the staircase*

   a) Ornately designed bannister
   b) Surly doorman
   c) Witty remark that occurs to you too late
   d) Shaggy-dog story

3. KRENTENKAKKER (Dutch)
   Literal translation: *raisin pooper*

   a) Miser
   b) Sheep
   c) Garibaldi biscuit lover
   d) Nun

4. ACABAR EM PIZZA (Brazilian Portuguese)
   Literal translation: *to end in pizza*

   a) To get flattened
   b) To end in a mess
   c) To end with no punishment
   d) To disappear

5. YOKO MESHI (Japanese)
   Literal translation: *a meal eaten sideways*

   a) Dinner on the sofa
   b) Awkwardness of speaking a foreign language
   c) Sexual intercourse
   d) Conversation with someone sitting next to you

6. SMULTRONSTÄLLE (Swedish)
   Literal translation: *wild strawberry place*

   a) Place or moment where you feel depressed
   b) Place or moment where you feel naughty
   c) Place or moment where you feel hungry
   d) Place or moment where you feel happy

7. LUFTMENSCH (Yiddish)
   Literal translation: *air person*

   a) Balloonist
   b) Uncharismatic person
   c) Impractical person
   d) Anesthetist

8. ATTACCABOTTONI (Italian)
   Literal translation: *button attacher*

   a) Overzealous cleaner
   b) Someone who won't leave you alone
   c) Tailor of couture
   d) Sponger

9. SEITENSPRUNG (German)
   Literal translation: *a jump to the side*

   a) Technique for placing one's beach towel
   b) Extramarital affair
   c) Non sequitur
   d) Overtaking maneuver

10. UITWAAIEN (Dutch)
    Literal translation: *to out blow*

    a) To walk in the wind for fun
    b) To fart in appreciation of a meal
    c) To enjoy a cannabis cigarette outdoors
    d) To shout cathartically when no one is around

10

Oh My Days

CALENDARS AND COMPASSES

We've traveled more than six thousand years, from the proto-writing of Sumer to languages invented since the turn of the millennium. We've been around the world, from Britain to Burma and from the Arctic to the Sahara. This chapter will take us on further adventures through time and space. We will diarize with the ancient Maya, sail across a windy Mediterranean, and get thoroughly lost on the slopes of a Melanesian volcano. We start with the Sami of northern Scandinavia.

Synchronize watches, it's problem o'clock.

SNOW TIME

The following clock times are written in Sami:

15:40	*Diibmu lea logi badjel beal njeallje*
16:50	*Diibmu lea logi váile vihtta*
13:10	*Diibmu lea logi badjel okta*
21:25	*Diibmu lea vihtta váile beal logi*
19:05	*Diibmu lea vihtta badjel čieža*
12:30	*Diibmu lea beal okta*

What are these times in Sami?

15:55   16:20   18:35   22:10

Despite the great linguistic diversity in the world, words for the days of the week are remarkably similar wherever you go. Most countries use one of two systems (or a combination of both): day names relate to planets and gods, or they are numbered from one to seven.

The Jews invented the "modern" seven-day week, in which seven days are repeated in an endless loop with no extra days to account for religious or astronomical adjustments. The Romans adopted the concept, and named the days after the Sun, the Moon, and the five planets visible to the naked eye. Most western European languages follow Latin's lead, although in Germanic languages, like English, Norse gods replace the planets. In the Slavic languages (and Portuguese), on the other hand, weekdays are numbered.

In China and Japan, the five visible planets are named after fire, water, wood, metal, and earth, the five elements of Taoist philosophy, which remain in the Japanese calendar.

A WEEK IN TOKYO

The Japanese days of the week, in order, are named after the Sun, the Moon, fire, water, wood, metal, and earth. Here are the English transliterations of some Japanese words, with their translations:

Nichibotsu	Sunset
Mokuhanga	Wood blockprint
Kaji	Fire event
Suimen	Water surface
Suion	Water temperature
Donabe	Earthen pot
Kin	Gold
Dochi	Land
Mokuyobi	Thursday
Getsumen	Moon's surface

Can you work out the Japanese words for the days of the week?

I've already included a problem in Swahili (see page 184), but I couldn't resist slipping in another. The names for the days of the week in Swahili are very different from those in European or Asian languages. The following problem gives you all the clues you need to work out how they differ.

93

TELLING THE TIME IN TANZANIA

Here are eight expressions of time in English and Swahili. The English terms are ordered chronologically, and the Swahili ones are ordered alphabetically.

1. Sunday, 1:00 AM

a) *jumamosi, saa moja usiku*

2. Sunday, 7:30 AM

b) *jumamosi, saa mbili na robu usiku*

3. Sunday, 9:15 AM

c) *jumamosi, saa nne na nusu asubuhi*

4. Tuesday, 12:15 PM

d) *jumamosi, saa saba usiku*

5. Tuesday, 11:30 PM

e) *jumanne, saa sita na robu asubuhi*

6. Saturday, 10:30 AM

f) *jumanne, saa tano na nusu usiku*

7. Saturday, 7:00 PM

g) *jumapili, saa moja na nusu asubuhi*

8. Saturday, 8:15 PM

h) *jumapili, saa tatu na robu asubuhi*

Match the English expression to the correct Swahili translation.

What day is *jumatatu* and why?

According to *The Hitchhiker's Guide to the Galaxy*, the answer to life, the universe, and everything is 42.

The Akan people of Ghana and Ivory Coast (whom we met on page 223) are probably the only people in the world for whom this statement is partially true. The traditional Akan calendar, or *adaduanan*, is 42 days long.

The *adaduanan* is a synthesis of two competing weekly cycles used by the Akan's ancestors: a six-day week and a seven-day week. Rather than choose between the two, the Akan run them both together. Six multiplied by seven equals 42, so this number is the length of one full cycle.

EVIL DAYS

The following table lists the names of days in the *adaduanan*, but it's only partially filled in. What are the names of the days marked a), b), c), and d), each nine days apart from one another, and which are known as "evil days," on which farming is forbidden?

	Week 1	Week 2	Week 3	Week 4	Week 5	Week 6
Mon	a)			Kurudwo		Monodwo
Tues			Kurubena			
Wed	Nkyiwukuo	b)		Monowukuo		
Thurs		Kwaya	Monoya			
Fri			c)		Nkyiafi	
Sat	Monomene					
Sun	Fokwasi			d)		Monokwasi

In previous chapters, I told the story of how the alphabet was originally invented in ancient Egypt and then spread north, via the Levant, evolving into the alphabets of Europe, the Middle East, and Asia. The alphabet also migrated southward, reaching the Horn of Africa, its geographic limit in that continent.

The Ge'ez alphabet is used by several languages in that region, including Amharic, the official language of Ethiopia. It is an alphasyllabary, like the Javanese and Lontara scripts I mentioned in the previous chapter, in which letters are consonants and diacritics are vowels. It's called Ge'ez because that's the name of the ancient Ethiopic language that first used it.

If you like reggae, you may recognize Ge'ez from band posters or album covers. Rastafarianism, the Jamaican religion and cultural movement, venerates Ethiopian history, and as a result Ge'ez has become part of reggae iconography.

As well as the alphabet, another ancient Egyptian cultural artifact traveled southward to Ethiopia: the calendrical practice of dividing the year into 13 months. Of these, 12 have exactly 30 days (giving 360 days), and the thirteenth is a mini-month, lasting only five or six. (This calendar also survives in modern Egypt, where it is used by farmers and members of the Coptic church.)

If you'd like to know the pronunciations of the unfamiliar letters shown below, see the Appendix. You don't need to know them to solve the problem.

95

A YEAR IN ADDIS ABABA

Here are the names of the 13 Ethiopian months in Amharic, along with their starting dates:

መስከረም	September 11 or 12	ሚያዝያ	April 9
ጥቅምት	October 11 or 12	ግንቦት	May 9
ኅዳር	November 10 or 11	ሰኔ	June 8
ታኅሣሥ	December 10 or 11	ሐምሌ	July 8
ጥር	January 9 or 10	ነሐሴ	August 7
የካቲት	February 8 or 9	ጷጉሜ	September 6
መጋቢት	March 10		

The transliterations of these words (listed in alphabetical order) are:

Gǝnbot	Mäskäräm	Säne	Yäkatit
Ḥamle	Miyazya	Taḫśaś	
Ḥǝdar	Nähase	Ṭǝqǝmt	
Mägabit	Ṗagume	Ṭǝrr	

Match the months to their transliterations.

Alongside Mesopotamia/Egypt and China, the third region where writing emerged independently was Central America. The ancient Maya, who lived in an area that now covers Guatemala, Belize, and parts of southern Mexico, had the most developed writing system of the dozen or so Mesoamerican civilizations with their own scripts. Mayan writing emerged during the "classical" Mayan period, between around 250 and 900 CE, and consists of more than 800 ornate and intricate glyphs, often depicting human or animal faces. Thousands of examples of the writing survive, carved in wood and stone, molded in stucco, and painted on ceramics and bark paper.

Mayan writing is a hybrid system, like Egyptian hieroglyphics, in which the symbols can represent the objects they are pictures of, or phonetic elements like letters or syllables.

The major work on deciphering the script was carried out in the 1970s and 1980s, and it is now estimated that we understand about 80 percent of the glyphs.

Mayan writing reveals that the Maya had the most sophisticated system of timekeeping of any ancient civilization. They used three calendars concurrently:

The Long Count. A cycle of 2,880,000 days (about 7,900 years). It was used only on monuments, and, extrapolating backward from dates found on these monuments, the clock would have started in 3114 BCE. The number 2,880,000 is $20 \times 20 \times 20 \times 18 \times 20$, since the Maya counted (mostly) in cycles of 20.

The Solar Year. This cycle was used by farmers, and was composed of 18 months of 20 days, and a nineteenth month of 5 days. The Maya ignored the extra quarter day and did not have leap years, so every four years the dates shifted out of sequence by a day.

The Tzolk'in. The religious calendar, the most important of the three, and the subject of the next puzzle.

In parts of the Guatemalan highlands, the *Tzolk'in* has been in continuous use since pre-Columbian times, surviving against the odds. Mayan culture has been brutally supressed for centuries, first by the Spanish and more recently in the Guatemalan civil war of 1960–1996, in which about 40,000 people of Mayan descent were killed. Freedom of worship was enshrined in the Peace Accords of 1996, and observance of the *Tzolk'in* is growing again, a visible way in which the country's six million Maya are reasserting their cultural identity. Every Mayan archaeological site now has a designated altar for traditional rituals.

96

TZOLK'IN ABOUT A REVOLUTION

The ceremonial *Tzolk'in* calendar of the ancient Maya is a 260-day cycle that repeats ad infinitum. In Mayan script, the dates of the *Tzolk'in* consist of two glyphs, one on the left and one on the right. All the left-hand glyphs repeat after the same number of days, and all the right-hand glyphs repeat after the same number of days. (These two numbers are different.)

Some *Tzolk'in* dates are shown on the next page, in a calendar for August and September 2007.

			1 AUGUST	2	3	4
5	6	7	8	9	10	11
12	13	14	15	16	17	18
19	20	21	22	23	24	25
26	27	28	29	30	31	1 SEPTEMBER
2	3	4	5	6	7	8
9	10	11	12	13	14	15
16	17	18	19	20	21	22
23	24	25	26	27	28 ?	29
30						

1. What is the correct glyph for September 28, 2007, marked "?" on the calendar?

2. Where on the calendar should the following dates fall?

a)   b)

It's time to move on. To a problem about movement. In Chinese, when an action involves motion, the words one uses to describe it depend on the direction of the motion relative to the position of the speaker.

97

THINKING INSIDE AND OUTSIDE THE BOX

The following diagrams are each described by a sentence in Chinese, written in *pinyin*, the standard way the language is written in the Latin alphabet. The black dot represents the speaker; the white dots are other people. The arrows depict the actions described in the sentences. I've also provided English translations of the first two sentences.

Translate the remaining sentences, and write a *pinyin* sentence that corresponds to the diagrams labeled a and b.

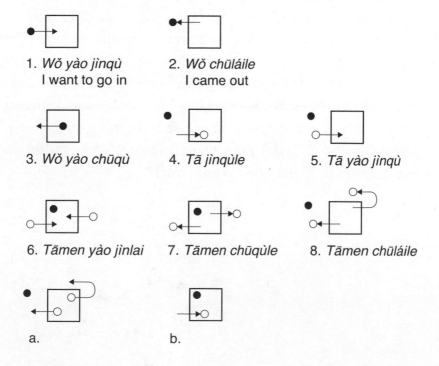

1. *Wǒ yào jìnqù*
   I want to go in

2. *Wǒ chūláile*
   I came out

3. *Wǒ yào chūqù*

4. *Tā jìnqùle*

5. *Tā yào jìnqù*

6. *Tāmen yào jìnlai*

7. *Tāmen chūqùle*

8. *Tāmen chūláile*

a.

b.

The cardinal compass directions are north, east, south, and west, and the four "intercardinals" are northeast, northwest, southeast, and southwest. In almost all languages, the words for the inter-cardinals are combinations of the words for the cardinals, as in English. In a tiny number of languages, however, such as Finnish, Estonian, and Maltese, the intercardinals have no etymological connection to the cardinals.

Finnish and Estonian belong to the Uralic language family, which also includes Sami (problem 91) and Hungarian (problem 99). I feel I've got the Uralics covered in this chapter.

Maltese, on the other hand, is a form of Arabic—the only Arabic language that is written in the Latin alphabet, and the only Arabic language that is an official language of the European Union. Malta is situated between Tunisia, Libya, and Sicily; Maltese vocabulary is mostly a mix of Arabic and Sicilian.

Now back to the compass. The Maltese words for seven of the eight compass directions originate from words for winds used in the Mediterranean Lingua Franca, a pidgin used for international commerce in the thirteenth and fourteenth centuries. Its vocab was made up mostly of words from Venetian and Sicilian, but also from Provençal, Catalan, Greek, and Arabic. Lingua Franca's use as a bridging language for people with different mother tongues gave rise to the modern meaning of the term. Marine compasses had Lingua Franca terms to denote wind directions, and in Malta, these names stuck.

98

A MALTESE TEASER

In Malta, the four cardinal and four intercardinal compass directions, listed alphabetically, are called:

Il-Grigal	Il-Lvant	In-Nofsinhar	It-Tramuntana
Il-Lbiċ	Il-Majjistral	Il-Punent	Ix-Xlokk

Maltese schoolchildren learn mnemonics to remember them, such as the two shown below:

Toni liebes nuccali pulit	Tony is wearing a smart pair of glasses
Gloria xammet laħam mixwi	Gloria smelled roast meat

Using the information in the statements below, deduce which Maltese names correspond to which compass directions.

1. Two directions are derived from the names of countries.
2. Two other directions—which are opposites—have Latin roots. One of these root words gave rise to the name of a region in the Middle East.
3. One other name with a Latin root means "a cold wind that comes from high land."
4. The Maltese word *nofsinhar* also means a time of the day.

For the next problem, we're back to the four cardinal directions.

99

HUNGARIANS IN A FIELD

The picture below shows a field divided into a 7 × 7 grid. North is at the top and east on the right. In some squares there are rocks, indicated by the black circle •. Four Hungarians— András, Béla, Csaba, and Dorottya—are standing in the field. Each is in a different square that does not contain a rock, and each is facing in one of the four cardinal directions.

Each person makes three statements describing the positions of the rocks from their standpoint. The first statement from each person describes a compass direction and, in brackets, how that direction relates to the way they are facing. The picture shows where András is, and the arrow points to the way András is facing. All directions describe single straight lines, such as "due east," or "directly behind me." I've given you the translation of András's first statement.

Find the positions of Béla, Csaba, and Dorottya, and the directions they are facing.

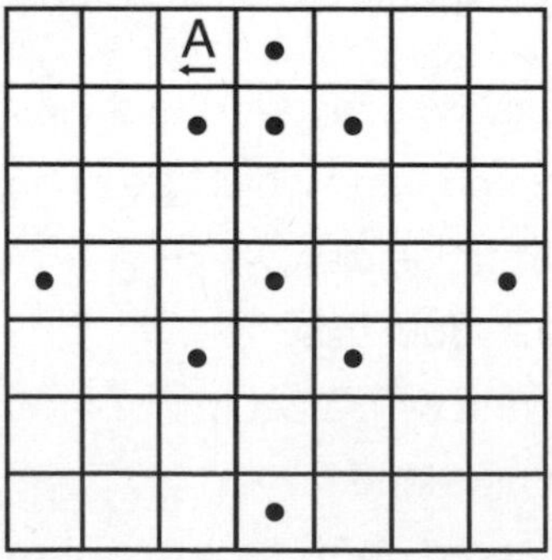

András says:

> Keletere (mögöttem) egy kő van   To the east (behind me)
> there is one stone
>
> Délre két kő van
> Jobbra nincs kő

Béla says:

> Délre (balra) nincs kő
> Északra egy kő van
> Mögöttem két kő van

Csaba says:

> Északra (előttem) nincs kő
> Nyugatra egy kő van
> Jobbra két kő van

Dorottya says:

> Nyugatra (jobbra) két kő van
> Északra egy kő van
> Balra nincs kő

And now for the grand finale. When I started researching this book I asked three of Britain's most successful contestants at linguistics olympiads, Sam Ahmed, Liam McKnight, and Elysia Warner, to give me a list of their favorite olympiad questions. Only two problems were on all three lists, and I am setting one of them here. (The other one is far too hard for this book.) It is the perfect olympiad problem. It transports you to a world you probably knew little about, in this case a small volcanic island off the coast of Papua New Guinea. It presents you with a perplexing puzzle. It reveals a fascinating feature of the local culture, which is interesting both linguistically and mathematically. And it sparks wonder, by revealing diverse ways in which different people approach even the most basic things.

In this case we learn that the inhabitants of the island do not use compass directions when describing relative positions. The method they do use is a simple one, even though it is challenging to deduce. I guarantee you will enjoy this problem, whether you peek at the answer or solve it by yourself.

The island is Manam, one of the world's most active volcanoes. Its ash clouds frequently rise a couple miles into the sky. Inhabitants are often evacuated. Located 8 miles (13 km) from the coast of Papua New Guinea, Manam is almost a perfect circle with a diameter of about 6 miles (10 km).

100

LANGUAGE LAVA

Here is a map of Manam Island, with seven houses marked on it. Five of the houses are labeled with their owners' names. The remaining two are labeled A and B. The + marks the highest point of the volcano.

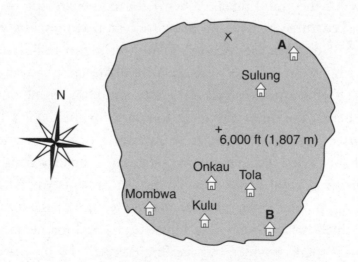

The following sentences in Manam describe the relative locations of the houses:

1. *Onkau pera kana auta ieno, Kulu pera kana ilau ieno.*

2. *Mombwa pera kana ata ieno, Kulu pera kana awa ieno.*

3. *Tola pera kana auta ieno, Sala pera kana ilau ieno.*

4. *Sala pera kana awa ieno, Mombwa pera kana ata ieno.*

5. *Sulung pera kana awa ieno, Tola pera kana ata ieno.*

6. *Pita pera kana ilau ieno, Sulung pera kana auta ieno.*

Who lives in A and B?

Arongo is building a house at X. Which statement is correct?

a) *Arongo pera kana awa ilau ieno, Sulung pera kana ata auta ieno.*

b) *Arongo pera kana ata auta ieno, Sulung pera kana awa ilau ieno.*

c) *Arongo pera kana ata ilau ieno, Sulung pera kana awa auta ieno.*

d) *Arongo pera kana awa auta ieno, Sulung pera kana ata ilau ieno.*

Remember, the Manam islanders *do not* use the terms "north," "east," "west," and "south." If you assumed they did, your mind would go around in circles.

ANSWERS

LINGO BINGO
Vocab Test A–L

1. d	5. c	9. c	13. d
2. b	6. d	10. d	14. a
3. d	7. c	11. c	
4. a	8. d	12. b	

1. Ok-Voon Ororok Sprok

1 ODD COUPLES

Here are some examples I found using Google:

a) "I did all that I **could to** please my God."
b) "Does **he have** it?"
c) "I can see **that that** will be a problem."
d) "I saw **the John** Lewis ad."
e) "Better **that than** the alternative."

2 ICE CHEESE

The word pair "puzzle tree" appears frequently as part of the common triplet "monkey puzzle tree."

credit	**card**	game		
ice	**cream**	cheese		
beach	**house**	prices		
couch	**potato**	chip		
bowling	**alley**	**cat**	food	
salmon	**fishing**	**boat**	trip	
space	**shuttle**	**bus**	**stop**	sign
cheese	**sandwich**	**shop**	**floor**	plan
eye	**contact**	**information**	technology	
cowboy	**boot**	**camp**	**fire**	department

3 IT STARTED TO RAIN

a) newspaper

instance	alternative
i)	print-media company
ii)	*Guardian* (or any other title)
iii)	pages

b) get

instance	alternative
i)	understand/recognize
ii)	am/become
iii)	obtain/gain
iv)	travel/go
v)	get/buy

c) good or great

instance	alternative
i)	opportune/right/appropriate
ii)	decent/skillful/trustworthy
iii)	loyal/dear/close
iv)	nice/fine/honest
v)	fun/enjoyable/happy

4 THE BAD TRANSLATION

incorrect translation	original word
façade	front
supply	arm
deceased	late
polished	shone
departs	leaves
ignite	light
confront	face
changed	turned
stabilize	steady

arrived	came
buying	getting
deceased	late
ahead	before
construct	make
house	home
thin	fine
departed	left
support	back
house	home

5   THE WORLD'S FUNNIEST CROSSWORD

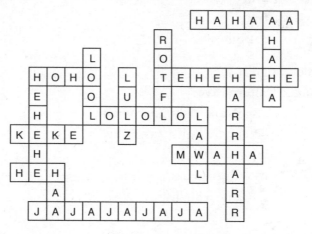

The given word LULZ reveals that another word must have an L three letters in. The only possible clue that fits is L(OL)+ and the word must be LOLOLOL. From this we know that one five-letter word must end in L. There are three candidates: LAWWL, LOOOL and ROTFL. However, we can eliminate LAWWL since if this were in the grid, one word must begin with a W, and none does. Likewise we can eliminate LOOOL, since no word begins with an O. Therefore the five-letter word ending in L must be ROTFL. Since there is only one clue that begins with a T, the eight-letter word beginning with a T must be TEHEHEHE or TEHEEHEE. We can eliminate the latter since no words begin with an E. I leave it to you to figure out how the other words fit.

6   WHO DO YOU THINK YOU ARE?

The Soundex algorithm works as follows:

Step 1: Keep the initial capital letter of the surname as the first letter of the code.

Step 2: Remove the letters "w" and "h."

Step 3: Replace the consonants with numbers using the table.

Step 4: Reduce any sequence of two or more identical digits to a single digit.

Step 5: Remove the vowels—"a," "e," "i," "o," "u"—and "y."

Step 6: If there are fewer than three numbers left, add zeros until there are three numbers.

Step 7: If there are more than three numbers left, keep only the first three numbers.

Ferguson	F622	Maxwell	M240
Fitzgerald	F326	Razey	R200
Hamnett	H530	Shaw	S000
Keefe	K100	Upfield	U143

7   WHAT'S MY NUMBER IN ITALY?

a) Pedro Santos, male, born on 3/2/86 in Brazil.

b) `ZGLLJN 82D46 Z100`

The *Codice Fiscale* is created in the following way:

Positions 1–3: The first, second, and third consonants of the surname. If there are not enough consonants, add the first vowel.

Positions 3–6: The first, third, and fourth consonants of the first name, if the name has four or more consonants. If the first name has only three consonants, the first, second, and third consonants are used. (If there are only two consonants, use both consonants and add the first vowel. If there is only one consonant, use the consonant and follow it with the first two vowels in the name. If the name has only two letters, the third space is an X.)

Positions 7–8: Year of birth.

Position 9: Month of birth, according to the following code:

A = January, B = February, C = March, D = April, E = May, H = June, L = July, M = August, P = September, R = October, S = November, T = December.

Position 10–11. Day of birth and gender. For men, the day of birth. For women, the day of birth plus 40.

Positions 12–15. Place of birth. (Italians born in Italy use a code for the municipality, or *comune,* they were born in, which consists of a letter and three numbers.)

Position 12: For foreigners, a Z is used.

Position 13: Continent of birth. 1 = Europe, 2 = Asia, 3 = Africa, 4 = North America, 6 = South America.

Positions 14–15: The country of birth within a continent. Countries are ordered alphabetically. The problem uses countries whose Italian names are similar to the English versions (Europe: 10 = France/*Francia*, 15 = Greece/*Grecia*, 33 = Switzerland/*Svizzera*; South America: 00 = Argentina/*Argentine*, 01 = Bolivia/*Bolivia*, 03 = Chile/*Cile*).

Albania/*Albania* is the first country in Europe when ordered alphabetically, so its two-digit code is 00. The most likely candidate for 02 in South America is Brazil/*Brasile*.

Position 16: Checksum.

8   EMBED WITH A LINGUIST

A. mathematics	B. mathematician	C. number
D. one	E. two	F. first
G. second	H. position	I. time

I solved this by building up my own semantic map of the words, and then seeing how it best matched the embedding. My way in was the words "one," "two," "first," and "second," which are connected semantically like this:

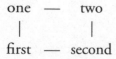

The word "time" is connected to the word "second"; likewise, the words "position" and "number" can be added onto the grid through their relationships to "first" and "second," and "one" and "two," respectively.

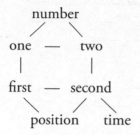

The words "mathematics" and "mathematician" clearly share some meaning, and it is likely that "mathematics" is also close to "number." By adding these two words to my mapping, and then rotating the whole thing 90 degrees counterclockwise, we can see how it fits neatly with the embedding.

9 A CROMULENT CONUNDRUM

Humans: B, C, E, F   Bots: D, G
Unclear: A, H, I

Each of the real statements contains two words. Our first task is to place the couplets of words on a scale of relative intensity. If a correct snippet is, for example, "cromulent but not melaxious," then "melaxious" is higher on the scale. Let's use the "<" symbol to denote "is less than" in terms of intensity. Then we can rewrite the 17 genuine samples as:

1. cromulent < melaxious
2. efrimious < quarmic
3. hyxilious < fligranish
4. daxic < fligranish
5. hyxilious < laxaraptic
6. melaxious < efrimious
7. quarmic < nistrotic
8. shtingly < efrimious
9. efrimious < tamacious
10. fligranish < optaxic
11. cromulent < shtingly
12. efrimious < nistrotic
13. tamacious < nistrotic
14. wilky < daxic
15. jaronic < daxic
16. jaronic < hyxilious
17. laxaraptic < optaxic

We can string these together in two networks, with the scale of intensity going from lower on the left to higher on the right.

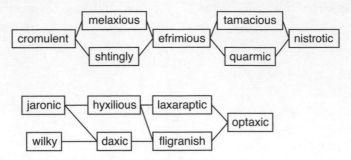

If both adjectives in an unknown snippet are in the same network, and the adjective described as higher-intensity is indeed higher in the diagram, then the snippet was written by a real person.

If both adjectives are in the same network but the adjective described as higher-intensity is lower in the diagram, then the snippet was written by a bot.

If each adjective is in a different network, then it is unclear who or what wrote the snippet.

For example, snippet A contains "hyxilious" and "quarmic." These words are in different networks, so it's unclear whether the author was human or bot.

Snippet B describes "jaronic" as less intense than "laxaraptic." Both words are in the same network, and "jaronic" is to the left of "laxaraptic," so this is a correct usage, and we can say with confidence that a human wrote it.

10 ■ WE COME IN PEACE

We have established that *farok* = *jjat*.

Progressing through the peace message, the second word is *crrrok*. It appears only in sentence 11, so our assumption is that it must be one of the Arcturan words *wat*, *nnat*, *arrat*, *mat*, or *zanzanat*. However, sentence 11 has six words in Centauri and only five in Arcturan. The Centauri words *crrrok* and *zanzanok* appear only in sentence 11, as does the Arcturan word *zanzanat*. Thus, *crrrok* and *zanzanok* must in some way correspond to *zanzanat.* Hmm. Let's move on. We'll come back to this later.

The third word is *hihok,* which appears in sentences 3, 11, and 12. By matching the Arcturan versions of these sentences, we deduce that *hihok = arrat,* since *arrat* only appears in these sentences. We also notice that *hihok* does not appear in the same place in the sentence as *arrat,* which confirms that the languages have a different word order.

The fourth word is *yorok,* which appears in 10 and 11. The only possible translation is *yorok = mat,* because the other words that repeat in 10 and 11 also occur in other sentences.

On to *clok.* This word appears only once, in 10.

10, Centauri: *lalok mok nok yorok ghirok clok.*

10, Arcturan: *wat nnat gat mat bat hilat.*

Clok cannot be *mat,* since that has been allocated already. We can also eliminate *wat, gat, nnat,* and *hilat,* since they appear in other sentences. (By matching, we can see that *lalok = wat, mok = gat, nok = nnat,* and *ghirok = hilat.*)

So *clok = bat.*

We go through a similar process to find *kantok* and *ok-yurp.* They both appear only in:

9, Centauri: *wiwok nok izok kantok ok-yurp.*

9, Arcturan: *totat nnat quat sloat at-yurp.*

The words *wiwok, nok, izok,* and *totat, nnat,* and *quat* all appear in other sentences; thus, we deduce that *kantok ok-yurp = sloat at-yurp.*

We are almost there. We have translated every word except *crrrok.*

Let's look again at 11, the only sentence in which *crrrok* appears.

11, Centauri: *lalok nok crrrok hihok yorok zanzanok.*

11, Arcturan: *wat nnat arrat mat zanzanat.*

As I argued above, *crrrok* and *zanzanok* must somehow correspond with *zanzanat.* What to do? It is most likely (but not certain) that *zanzanok = zanzanat* because they look similar and are at the same place in the sentence. So what does *crrrok* translate to? Well, maybe nothing. Given no indication of what else we are to do, we can ignore it in our translation.

Centauri peace message: *farok crrrok hihok yorok clok kantok ok-yurp*
Arcturan translation: *jjat arrat mat bat sloat at-yurp*

Congratulations, you have just translated between English and Spanish! The sentences listed in the puzzle are word-for-word copies (in Centauri and Arcturan) of these corporate phrases.

1. García and associates
   García y asociados
2. Carlos García has three associates
   Carlos García tiene tres asociados
3. his associates are not strong
   sus asociados no son fuertes
4. García has a company also
   García también tiene una empresa
5. its clients are angry
   sus clientes están enfadados
6. the associates are also angry
   los asociados también están enfadados
7. the clients and the associates are enemies
   los clients y los asociados son enemigos
8. the company has three groups
   la empresa tiene tres grupos
9. its groups are in Europe
   sus grupos están en Europa
10. the modern groups sell strong pharmaceuticals
    los grupos modernos venden medicinas fuertes
11. the groups do not sell zanzanine
    los grupos no venden zanzanina
12. the small groups are not modern
    los grupos pequeños no son modernos

The message of peace was:

Clients do not sell pharmaceuticals in Europe.

Which we translated to:

Clientes no venden medicinas en Europa.

LINGO BINGO

Loan Words

1. d) Greenlandic: *anoraq*
2. b) Ilocano: *yóyo*
3. c) Hindi: *chāmpo*, meaning "press!"
4. a) Hawaiian: *'ukulele*, meaning "jumping flea"
5. c) Spanish: diminutive of *mosca*, meaning "fly"
6. b) Greek: means "the many"
7. a) Samoan: *tatau*
8. b) Dutch: *taptoe!*, meaning "close the tap (of the cask)!"
9. c) Korean: 태권도, *tae-kwon-do*, meaning "foot-fist-way"
10. d) Ojibwa: *nindoodem*
11. b) Nahuatl: *tomatl*
12. d) Temne: *k'ola*, from the kola nut of the kola tree

2. Celts, Counts, and Coats

11 OGHAM, SWEET OGHAM

birch: *bethe*
spruce: *ailm*

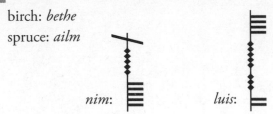

nim: *luis*:

As we learned in the introduction to this puzzle, each group of notches or strokes represents an individual letter. More precisely, notches are vowels and strokes are consonants. The words read from the bottom to the top. All the letters you need for the answers are included in the question, apart from the vowel "a," which you need to deduce. If "o" is two notches, "u" is three notches, "e" is four, and "i" is five, then by a process of elimination "a" must be the single notch.

12 IN THE STREET, HE SAW A MUTANT!

1. b   2. a   3. a   4. b

13 TOPONYM O' THE MORNING

1. Mullaghbane   *An Mullach Bán*   The White Summit
2. Knocknakillardy *Cnoc na Cille/Coille Airde* Hill of the High
                                                Church/Wood
3. Gortnabinna   *Gort na Binne*   Field of the Peak
4. Blackcastle   *An Caisleán Dubh*

The table of nouns that I hoped you would draw is illustrated on the next page. Position 1 is the first noun when there are two nouns in the name (and the basic form). Position 2 lists instances where there is a single noun in the name—in which case it always follows *An*. Position 3 is the "of" position, which is also the second noun when there are two in the name. I have also included the basic form of *Páirc*, which I gave you in the puzzle's setup.

I mentioned that the nouns fall into two categories that behave differently: those that have an "i" as a final vowel in Position 1, and those that don't. To see clearly what's going on, the former are underlined in the table below, the latter are not.

position 1	position 2	*an/na*	position 3	meaning
Cill		*na*	*Cille*	Church
Coill	*(An) Choill*	*na*	*Coille*	Wood
Páirc	*(An) Pháirc*	*na*	*Páirce*	Park
	(An) Currach	*an*	*Churraigh*	Marsh
Gort	*(An) Gort*	*an*	*Ghoirt*	Field
Baile				Town
Cluain				Meadow
Gleann				Valley
Talamh				Land
Bun				Base
	(An) Bhinn			Peak
	(An) Mhainistir			Abbey
	(An) Dún			Fort
		an	*Chnoic*	Hill
		an	*Chaisleáin*	Castle
		an	*Chairn*	Mound
		an	*Mhullaigh*	Summit
		na	*Muice*	Pig

Here's the pattern. Nouns with a final "i" in position 1 behave as follows:

In position 2 they gain an "h" before the first vowel.

They use the article *na* as the "of" form.

In position 3 they gain the suffix "-e."

The other nouns:

Have the same form in positions 1 and 2.

Use the article *an* as the "of" form.

In position 3, they gain an "h" after the first consonant, and an "i" before the last consonant cluster. (And sometimes the "c" is changed to a "g" in that final cluster.)

The adjectives, for their part, agree with the nouns, and reflect both their position in the sentence and the class of the noun.

For example, for nouns with an "i" in the final position (underlined below), the adjectives gain an "h" in position 2, and an "i" and an "e" in position 3. Nouns that do not have an "i" as a final vowel (not underlined below) keep their basic form in position 2, and gain an "h" and an "i" in position 3.

position 1	position 2	an/na	position 3	meaning
	(Mhainistir) Dhubh	na	(Muice) Duibhe	Black
	(Gort) Bán	an	(Ghoirt) Bháin	White
	(Bhinn) Bhán			White
	(Dún) Ard			High

Now to the four questions.

1. Mullaghbane will be "The White Summit," in the form *An* ["Summit"] ["White"].

By seeing that *Currach* in position 2 becomes *Churraigh* in position 3, we deduce that *Mhullaigh* in position 3 becomes *Mullach* in position 2. We know that nouns in position 2 that do not have "i" as the final vowel take the adjective in its basic form, in this case *bán*. So the answer is *An Mullach Bán*.

2. Knocknakillardy will be "Hill of the High Wood/Church," written ["Hill"] *na* ["Wood/Church"] ["High"].

We know that *Chnoic* ("Hill") in position 3 takes the *an* form for "of," and thus in its basic form does not have an "i" as a final vowel. We must thus extract the "h" after the first consonant and the "i" before the last consonant cluster, which gives us *Cnoc* when in position 1. We know that *Cille/Coille* are the correct forms for position 3. Since the adjective agrees with the noun, *Ard* must also add an "i" and an "e," as in the observation above. So we can deduce that the answer is *Cnoc na Cille/Coille Airde*.

3. Gortnabinna will be "Field of the Peak," written ["Field"] *na* ["Peak"]. "Field" in position 1 is *Gor*. We know that "Peak" in position

2 is *Bhinn* (i.e., last vowel an "i"), and thus in position 3 must take the *na* form of "of" and mutate to *Binne*. The answer is *Gort na Binne*.

4. Blackcastle will be *An* ["Castle"] ["Black"]. Since "Castle" in position 3 is *Chaisleáin*, we extract the "h" and the "i" to place it in position 2. We know that for nouns that have an "i" in their final vowel, the position 2 form of "black" is *dhubh*. Thus, for nouns that do not have an "i" in their final vowel, the position 2 form is simply *dubh*. So the answer is *An Caisleán Dubh*.

Congratulations if you made it this far. Pints of Guinness all around!

14 WE ALL LOVED THE GIRL

1. The king loved you all
2. You all loved the girl
3. We two loved you two
4. *se æpeling lufode þæt cild*
5. *þæt cild lufode þone æpeling*
6. *we lufodon þæt cild*
7. *þæt cild inc lufode*

Old English has several grammatical features that no longer exist in Modern English.

Verb endings: *lufode* is the third person singular form of the verb "to love," and *lufodon* is the form used for plurals.

Gender: Old English has three genders: masculine, feminine, and neuter. (Only masculine and neuter nouns are used in this problem.)

Personal pronouns: Old English distinguishes between "we (the two of us)" and "we (all of us)," and between "you (the two of you)" and "you (all of you)." Modern English retains this distinction in "both" versus "all."

Cases: The words for "we," "you," and "the" change according to whether they are the subject or the object of the sentence. For masculine nouns like "king" (*cyning*) and "prince" (*æpeling*), "the" = *se* when it is the subject and "the" = *þone* when it is the object. For neuter nouns like "girl" (*mægden*) and "child" (*cild*), "the" = *þæt* for both subject and object.

Word order: Because of verb endings and cases, word order does not matter so much in Old English, since the sense can be deduced from the form of the words used.

15 ▍ IT'S GRIMM UP NORTHWEST EUROPE

Proto-Germanic	English	German	Icelandic
*krampaz	cramp	Krampf	krampar
*aplu	**apple**	***Apfel***	epli
*swanaz	swan	Schwan	svanur
*þrīz	three	drei	þrír
*swīnan	**swine**	***Schwein***	**svír**
*jæran	year	Jahr	ár
*þūman	**thumb**	Daumen	þumalfingur
*þurnuz	**thorn**	***Dorn***	þyrnir
*wurðan	**word**	Wort	orð
swerðan	sword	Schwert	sverð

(Note: *Apfel* is the German word, but if you got *Apfle* give yourself a point. Given the material in the problem, it's not possible to work out which one is correct.)

Here's how you may have started: If *swanaz* becomes *swan/Schwan/svanur*, then we deduce that *sw-* becomes *sw-/Sch-/sv-*. Thus we can fill the first two letters of each translation of *swinan*: *sw_ _ _*, Sch _ _ _ _, *sv _ _*, and we can fill in the first two letters of the Proto-Germanic for sword: *sw _ _ _ _ _*. The letter "*þ-*" seems to become *th-/D-/þ-*, so we can fill in those too.

Now look at the ends of the words. If *Wort/orð* evolved from *wurðan*, then it seems likely that *Schwert/sverð* come from a word ending *-rðan*. And if the English for *Schwert/sverð* is "sword," then we can be fairly sure that the English for *Wort/orð* ends with "-rd." If you get stuck with the English for *þūman*, you will be happy that the Icelandic gives you the "-fingur."

16 TRIPLE DUTCH

vliegen: to fly, flies (insects)

weg: road, away

bij: bee, with, near to

graven: to dig, grave, counts (noblemen)

17 AN ARMOIRE OF COATS

1. f	3. h	5. g	7. b	9. c
2. d	4. i	6. a	8. e	

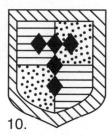

10. 11.

Blazon descriptions have no punctuation, and they start with the background. Each feature is followed by its color, which has an initial capital. Arrangements of objects are given row by row. Objects on the left are *sinister* and on the right *dexter*, the Latin terms for left and right, although this is from the shield-bearer's perspective, so the opposite of what's on the page.

A glossary of the phrases used in this problem:

Colors

Argent	white/silver
Azure	blue
Gules	red (the derivation is believed to be from the old French for "throat," possibly because neckpieces were made from red fur)
Or	gold/yellow
Sable	black (from the sable, a marten with black fur)
Vert	green

Positions (remember that left and right are used here from the shield-bearer's perspective)

per pale	in vertical halves—describe the left half first (This can also be described by X *impaling* Y, when X is on the left and Y on the right)
per chevron	inverted V shape—describe the top half first
per fess	horizontal halves—describe the top half first
per bend	diagonal sloping to the left—describe the bottom left first
quarterly	in quarters, in this order: top left, top right, bottom left, bottom right
chief	in the top part
counterchanged	an object superimposed on a split background, and which is filled with the opposing colors
overall	in the middle

Features

annulet	ring
bend	diagonal band sloping to the left
bend sinister	diagonal band sloping to the right
bordure	border
chequy	checkered
cross crosslet	crossed cross
embattled	with square teeth
escutcheon	shield
fess	horizontal band
goutte	droplet
lozenge	diamond
rose	rose
roundel	circle

18 IT'S ALL GREEK TO ME

Bibliophobia	fear of books
Cardialgia	heart pain
Dendrolatry	the worship of trees

Dromomania	a mania for roaming or wandering. From *dromos* ("course," "running")
Gynophilia	love of women
Hippophobia	fear of horses
Misandry	hatred of men
Misanthropy	hatred of humans
Misogamy	hatred of marriage
Misopedia	hatred of children
Monandry	the practice of having only one husband at a time
Morosoph	a wise fool
Mystagogue	a mystical leader. From *-agogos* ("leader," "guide")
Pelotherapy	the application of mud as a therapeutic treatment
Philanthropism	the desire to help others, especially by giving money to good causes
Photagogue	a leader who brings enlightenment
Polydactyl	a person with more than five fingers or toes on their hands or feet
Telesthesia	the supposed perception of distant occurrences or objects by means other than the known senses

19 SURE, SURE, MR. SHAW

1. b   2. d   3. e   4. a   5. c

ʃ is the sound "ss" and ʒ is the sound "zz"
ɟ is the sound "f" and ſ is the sound "v"

In other words, each pair of characters (which is the same character rotated by 180 degrees) shows the "unvoiced" and "voiced" version of the same sound. These terms refer to how the mouth pronounces the sound: The unvoiced version does not vibrate the vocal chords, whereas the voiced one does.

The sound "b" is a voiced consonant, and its voiceless counterpart is "p." Since the Shavian character for "p" is), we can deduce that the character for "b" must be the symbol) rotated 180 degrees, which is: (.

20 WARI BILONG YU

leg of dog	*lek bilong dok*
saltwater	*salwara*
grass	*gras*
bed	*bet*

haus bilong yu	your house
haus bilong king	palace
haus bilong wasim klos	laundrette
haus dok sik	animal hospital/vet
haiskul	high school
gras bilong het	hair (on head)
gras bilong fes	beard
maus gras	mustache
katim gras (two meanings)	cut grass/haircut
pen bilong maus	lipstick
klos meri	women's clothes
tekewe klos	undress
bret	bread
kukim bret	bake bread
kikbal	football
susok man	person who wears shoes and socks—i.e., someone who lives in a town

LINGO BINGO

Polish Phonetic Spelling

1. *Apmynste*: Upminster
2. *Datfed*: Dartford
3. *Douwe*: Dover
4. *Dzylynem*: Gillingham
5. *Eszfed*: Ashford
6. *Hejstynz*: Hastings
7. *Hen-Bei*: Herne Bay
8. *Istbon*: Eastbourne
9. *Koulczyste*: Colchester
10. *Luys*: Lewes
11. *Łytstebl*: Whitstable
12. *Magyt*: Margate
13. *Njuhejwn*: Newhaven
14. *Saufend-on-Sji*: Southend-on-Sea

3. All About That Base

21 AYE, IAAI

7, 9, 11, 13, 15, 17, 19

The repetition of *ke nua* leads us to consider that this probably means "and," connecting the parts before and the parts after. Since the numbers are 2 apart, the penultimate line is 10 more than the first line, and has the same phrase with the extra word *libenyita*. Likewise, the last line is 10 more than the second line, and also has an added *libenyita*. So it's likely that *libenyita* means 10. If it does, then *thabung* is less than 10. In fact, since *libenyita ke nua khasa* is 10 "and something," and *thabung ke nua vak* is less than 10 (if it was more than 10 it would include the word *libenyita*), we can deduce that *thabung ke nua vak* = 9 and *libenyita ke nua khasa* = 11. *Thabung* = 5, and the forms of the words are 5 + 2, 5 + 4, 10 + 1, 10 + 3, 10 + 5, 10 + 5 + 2, 10 + 5 + 4.

22 THE KNIGHTS WHO SAY NI

seks = 6
nioghalvtreds = 59
treogtyve = 23
femoghalvfems = 95
toogtres = 62
halvfjerds = 70

7 = *syv*
21 = *enogtyve*
54 = *fireoghalvtreds*
85 = *femogfirs*
99 = *nioghalvfems*

Here's how you might have worked it out. If *ni* = 9 and *nioghalvfjerds* = 79, it seems fair to assume that *-oghalvfjerds* describes "and 70." We see similar constructions with *og* among the other terms, so a reasonable guess would be that *og* is "and," and that the remaining part of the word denotes the multiple of ten.

That would make: *tyve* = 20, *halvtreds* = 50, *tres* = 60, and *firs* = 80.

The fact that the word for 60 looks like the word for 3, and the word for 80 looks like the word for 4, leads to the conclusion that this is a base-20 system. *Firs* = 4 × 20 (four score) and *tres* = 3 × 20 (three score). But why is *halvtreds* = 50?

The "odd" tens—50, 70, and 90—are considered "halfway" from one score to the next. Thus *halvtreds* means "halfway to the third score," as in "halfway between 40 and 60." Correspondingly, *halvfjerds* is "halfway to the fourth score," or 70, and *halvfems* is 90, or "halfway to the fifth score."

It's pretty confusing, especially for foreign tourists in Copenhagen, that the word for 50 doesn't have the word for 5 in it, but instead the words for "half" and 3. Indeed the Danish authorities once tried to introduce regular forms for numbers—*femti* (50), *seksti* (60), *syvti* (70), *otti* (80), and *niti* (90)—for bank notes and checks, but the words never caught on. In 1952, a 50-kroner banknote was introduced that said *femti kroner* on it. But no one ever said *femti*, so it was taken out of circulation, and in 2009 was replaced by one reading *halvtreds*.

23 MATH BABYLON

 a) 122　　　　　　　 b) 11　　　　　　　 c) 549

$$\mathsf{Y} = 1 \qquad\qquad \blacktriangleleft = 10$$

The Babylonian system uses two symbols, above, and two methods to express the value of a number.

First, when symbols are placed together you add them up, like Roman numerals, so $\mathsf{YY} = 2$, $\mathsf{YYY} = 3$, and so on. You can combine them together too, so $\blacktriangleleft\mathsf{Y} = 11$.

Yet the Babylonian system also has a positional element with a base of 60. The symbols in the rightmost group represent units, and the symbols in the group to the left represent multiples of 60.

Here are three numbers taken from the first part of the question, and how their values are expressed:

$(6 × 60) + (10 + 10 + 6) = 386$

$60 + (10 + 1) = 71$

$(10 × 60) + (50 + 4) = 654$

Thus:

 a) $(2 × 60) + (2 × 1) = 122$
 b) $10 + 1 = 11$
 c) $(9 × 60) + (9 × 1) = 549$

24 A LOAD OF OLD ROPE

The *khipu* system works as follows: Each string has a group of knots in three positions along the string. The number of knots in each position represents a decimal digit. The "o" knots encode the units digits, and the "x" knots encode the tens and the hundreds digits. The image below shows the number of knots in each position.

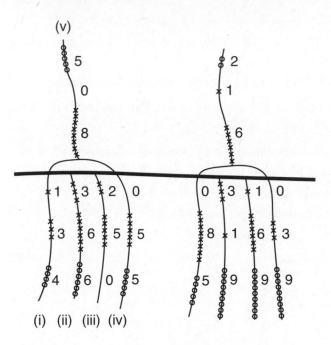

The strings marked (i), (ii), (iii), (iv), and (v) thus represent the numbers 134, 366, 250, 055, and 805. The number on the top string, which groups the four bottom strings together, is equal to the sum of the four bottom strings, so 134 + 366 + 250 + 055 = 805. We can confirm this with the set on the right. The bottom strings again add up to the top one: 085 + 319 + 169 + 039 = 612.

Now to the puzzle itself. The four bottom strings on the second image contain the numbers 089, 258, 273, and 038. These numbers add up to 658, so the top string must encode 658:

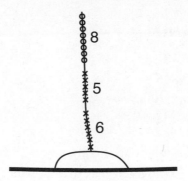

How might you have made this decipherment? Well, I did ask you to count the knots. The maximum number of knots per position is nine. In our decimal system, the maximum value for a digit is 9. It would be a fair assumption to suppose that the number of knots in each position denotes a decimal digit. You also knew that each string represents a three-digit number. There are clearly three groups of knots per string, so a logical approach would be to test the theory that each group of knots represents a single digit.

But why two types of knot? If a string represents a number that can be expressed in several digits, it is natural to assume that there must be a way to denote where the number starts (or finishes). The "o" knot performs this function: It is the unit position, and the other positions are the ascending powers of ten.

Extra puzzle: In the original edition, I wrote that, excluding interjections and exclamations, the best is MID = 1,499. A few readers got in touch to tell me that in the standard form of Roman notation, an I can only precede a V or an X, which means that MIX = 1,009 wins. Otherwise, some dictionaries include MMM (3,000) and MM (2,000) as words. Hmmm.

25 NANU NANU

kakiku	10,101
tajidu	180,816
nanu	200,020
bha	24

The first pattern you might spot is that *ka* = 1, *ki* = 100, and *kaki* = 101, so *ka* + *ki* = *kaki*, which tells us that when syllables are placed next to each other they are added together. (This is just like Roman numerals, since II = I + I, but unlike Arabic numerals, because 11 does not equal 1 + 1.)

We spot the syllable pair *kaki* in *kakigu* (30,101). We know that *kaki* = 101. Working on the presumption that syllables placed next to each other add up, we can deduce that *gu* must be 30,000. We could continue in this way, noticing that if *ga* is 3, and *gakiku* is 10,103, then *kiku* must be 10,100. And so on.

However, by now you may have spotted another pattern, which is how we will complete our understanding of Aryabhata's system:

- The first column contains one- and two-digit numbers, and all correspond with single syllables that end in an -*a*.
- The second column contains three- and four-digit numbers, and the final syllable is always -*i*.
- The third column contains five- and six-digit numbers, and the final syllable is always -*u*.

In other words, the vowel sound in a syllable indicates the number of decimal digits in the number, while the consonant sound indicates a two-digit value. We can see this most clearly with *da* and *du*, which are 18 and 180,000. If you keep the consonant, but change the vowel, the same two-digit number moves to a new position.

To be more precise, a syllable with an -*a* at the end determines the positions of the units and tens. (Sometimes, as in *ka*, *kha*, and *ga*, the tens position is 0.)

A syllable with an -*i* at the end determines the positions of the hundreds and thousands.

A syllable with a -*u* at the end determines the positions of the ten-thousands and hundred-thousands.

What makes this especially confusing is that Aryabhata's notation places the syllables representing units and tens on the left, rather than the right.

If the consonants indicate two-digit numbers, and the vowels indicate the decimal positions, we can fill out a table of the consonants and vowels that appear in the question. Note that not all of these syllables appear in the question: In some cases, such as, say, *gi*, we have deduced its meaning because we know *ga* is 3, and swapping an *a* for an *i* multiplies the number by 100.

1	*ka*	100	*ki*	10,000	*ku*
2	*kha*	200	*khi*	20,000	*khu*
3	*ga*	300	*gi*	30,000	*gu*
4	*gha*	400	*ghi*	40,000	*ghu*
8	*ja*	800	*ji*	80,000	*ju*
16	*ta*	1,600	*ti*	160,000	*tu*
18	*da*	1,800	*di*	180,000	*du*
19	*dha*	1,900	*dhi*	190,000	*dhu*
20	*na*	2,000	*ni*	200,000	*nu*
21	*pa*	2,100	*pi*	200,000	*pu*
23	*ba*	2,300	*bi*	200,000	*bu*

Thus *kakiku* = *ka* + *ki* + *ku* = 1 + 100 + 10,000 = 10,101

tajidu = *ta* + *ji* + *du* =   16 + 800 + 180,000 = 180,816

nanu = *na* + *nu* = 20 + 200,000 = 200,020

Which leaves us with *bha*. The consonant *bh*, the aspirated version of *b*, does not appear in the question. However, we see from *ka*/*kha*, *ga*/*gha*, and *da*/*dha* that the aspirated version of a consonant adds 1 to the value of the original consonant. Thus, it is fair to assume that *bha* = *ba* + 1 = 23 + 1 = 24.

26 MONKY PUZZLE

8,529

Cistercian numerical notation describes four-digit numbers (including one-digit, two-digit, and three-digit numbers with initial zeros). Each digit in a four-digit number occupies one of the four quadrants illustrated below. The vertical line is the stem that appears in all digits.

10s	1s
1,000s	100s

Thus, the units digit is drawn in the top right quadrant, the tens digit in the top left, and so on.

The numbers from 0000 to 0009 are shown below. The same (mirrored) patterns of lines in the top-left quadrant denote the tens digits. Likewise, the hundreds are shown in the bottom right and the thousands in the bottom left.

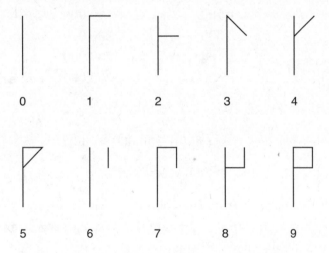

To solve the problem, you had to make the link that each quadrant represents a digit. In the first and the third Cistercian numbers, two quadrants have repeated patterns. These two numbers must be 1,410 and 5,750 (or vice versa) since these numbers have repeated digits. The repeated pattern on the first Cistercian number is a perpendicular line, while the repeated pattern on the third is a line and a diagonal. It would make more sense if the simplest one were 1, so a reasonable assumption is that the first number is 1,410. If it is, we can deduce the most likely ordering of the quadrants, and thus we know the patterns for 0, 1, 4, 5, and 7. Looking at the two remaining numbers, we can deduce that the second is 4,173 and the fourth is 1,368, which gives us the patterns for 3, 6, and 8.

The unknown number presented in the fifth symbol contains an 8 and a 5, and two other patterns we haven't seen before, which must be 2 and 9. Looking at the patterns for 0 and 1, and the progression of the patterns from 3 to 8 shown, we deduce that it's more likely that the pattern with a single line is 2, and the symbol with three lines is 9. Thus the answer is 8,529.

27 DOWN AND ACROSS IN ANTANANARIVO

We can solve this puzzle in a number of steps.

Step 1. The single words in the left-hand column contain the numbers from 1 to 10. One of these numbers, *folo*, is also an across clue. Since across clues represent numbers that are either two or five digits long, we can deduce that *folo* = 10.

Step 2. The number in *sivy*-across is 10. So *sivy* must be either 1, 3, 9, or 10, since these are the numbers with two-digit clues. It can't be 1 or 3, since, in either case, that would mean a down clue must have a zero as a first digit, which is forbidden. And it can't be 10, since we know *folo* is 10. So *sivy* = 9, and we can fill in the first batch of cells.

Step 3. Only the numbers 1 and 3 have both across and down clues. So *iray* and *telo* must be 1 and 3 (or vice versa). Thus *dimy*, *fito*, and *valo* are 5, 7, and 8 (but not necessarily in that order).

Step 4. In fact, we could have worked this out by looking at the number phrases. In the across clues, *dimy*, *fito*, and *valo* have number phrases that are considerably longer than the others. It would seem reasonable to presume that these represent five-digit numbers, while the three shorter phrases represent the two-digit numbers. Looking at the long number phrases, we can see that they are usually divided into five sections, connected either by *amby/ambin'ny* or *sy*.

They usually have the form:

[a number from 1 to 10] + [*amby* or *ambin'ny*] + [either *folo*, or a word that ends in *-polo*] + [*sy*] + [*zato* or a word that ends in *-jato*] + [*sy*] + [a number from 1 to 10 and *arivo*, or just *arivo*] + [*sy*] + [a number from 1 to 10 and *alina*, or just *alina*].

It would seem probable that each of these five sections describes a digit in the five-digit number. The question now becomes: Which section describes which digit?

Step 5. This is where we work out that the first section is the units, the second section the tens, the third section the hundreds, and so on. Consider the clue for *telo* across, which is *fito ambin'ny folo*, or *fito* + 10. In other words, the number is phrased such that the unit comes before the ten. The other two across clues that represent two-digit numbers (*iraika amby fitopolo* and *fito amby fitopolo*) do not contain any *-jato*, *arivo*, or *alina* phrases, which suggests that these terms are for numbers that are bigger than 100. Since these terms always come after the units and tens, our suspicions are confirmed that the numbers are spelled out in increasing powers of ten.

Step 6. The units digit of the clue for 1-down is 1. (We know this, since we have already filled a 1 in the bottom cell of the first column.) Which, according to the above rule, means that the first word of the 1-down clue is 1. Since either *iray* or *telo* is 1, the first word in the 1-down clue is either *iraika* or *fito*. The version that makes most sense is that *iray* and *iraika* are both words for one. So *iray* = 1, and *telo* = 3. We can now fill in one more cell. The first word in the 10-across clue is *iraika*, so we know that the units digit of that clue is a 1.

The 1 in this cell is also the units digit for 4-down. There is only one other unsolved down clue with *iraika* in the units position, so *efrata* = 4.

1	2		3	4
5		6		
7				
8				
9 1	0		10	1

Step 7. Only one of the down clues, 6-down, is a three-digit number. Looking at the number phrases, we can deduce that *enina* must be 6 and that the first word in the clue, *sivy*, or 9, must be in the units position. That leaves one down clue: 2-down. Looking at the down clues, the only number unaccounted for is *roa*. So *roa* = 2.

1	2		3	4
5		6		
7				
8		9		
9 1	0		10	1

Step 8. The hundreds word of *valo*-across is *sivinjato*, which suggests that *sivy*, or 9, is in the hundreds position of this number. No other clues contain a *sivinjato*, which suggests that *valo*-across is 8-across, and this coincides with the 9 already in the grid. So *valo* = 8.

Step 9. The hundreds word of *dimy*-across is *telonjato*, which suggests that *telo*, or 3, is in the hundreds position. We also know that *telonjato* is the hundreds word of 6-down, so this number begins with a 3. No other across clues have a *telonjato*, so *dimy*-across must be 5-across, and *dimy* = 5. By a process of elimination, *fito* = 7.

We can now fill in the units cells of all the across clues:

¹	²7	■	³	⁴7
⁵		⁶		4
⁷				5
⁸		9		8
⁹1	0	■	¹⁰	1

Step 10. We now get a better sense of the system. The tens digit is *folo* or, for any multiple of 10, it is the word corresponding to the multiple followed by the suffix *-polo*. The hundreds digit is similar, except with *zato* (meaning 100) as the base word, which becomes the suffix *jato*. Last, the thousands and ten-thousands places are denoted by [number] *arivo* or [number] *alina*, where [number] is omitted if the digit is 1. The connector *ambin'ny* is used between the ones and the tens places if the tens place is 1; otherwise the connector *amby* is used between the ones and tens places. Finally, the connector *sy* is used between all other decimal places. So the full grid is:

¹7	²7	■	³1	⁴7
⁵9	0	⁶3	6	4
⁷7	1	2	1	5
⁸1	5	9	6	8
⁹1	0	■	¹⁰7	1

28 PADDLING THE PACIFIC

We know that the vowels are the same, and that the consonants change in a systematic way.

Below I have rewritten the table in terms of the consonants in each word and when they appear. The two consonants in the words for "one" are in columns (i) and (ii). The consonants in the words for "two" are in column (iii), and so on.

i	ii	iii	iv	v	vi	vii	viii	ix	x	xi	xii	xiii	xiv
ONE		TWO	THREE		FOUR	FIVE		SIX	SEVEN		EIGHT		NINE
k	*h*	*l*			*h*	*l*	*m*	*n*	*h*	*k*	*w*	*l*	
t	*h*	*r*	*t*	*r*	*wh*			*n*	*wh*	*t*	*w*	*r*	*w*
t	*h*		*t*	*'*	*h*			*n*			*v*	*'*	
t	*'*			*'*		*r*	*m*	*n*	*'*	*t*	*v*	*r*	*v*
t	*s*	*l*				*l*	*m*	*n*	*f*	*t*			*v*

To solve the puzzle, we'll fill in the table, and then deduce the missing words.

From column (i) we can fill in columns (iv) and (xi).

Columns (iii), (v), (vii), and (xiii) coincide perfectly, as do columns (xii) and (xiv).

In addition, (vi) and (x) share enough consonants to be the same. When first looking at this problem, it might be confusing that *h* in Hawaiian is sometimes *h* in Maori (column (ii)) and sometimes *wh* (columns (vi) and (x)). Our new table makes it clear that there are two rules: one that applies when *h* is at the beginning of a word, and one that applies when it is in the middle.

To fill in column (viii), we can see from column (ix)—the only other case where Hawaiian, Rarotongan, and Samoan have the same consonants—that Maori and Nuku Hiva do, too.

The final table is printed on the following page.

i	ii	iii	iv	v	vi	vii	viii	ix	x	xi	xii	xiii	xiv
ONE		TWO	THREE		FOUR	FIVE		SIX	SEVEN		EIGHT		NINE
k	h	l	k	l	h	l	m	n	h	k	w	l	w
t	h	r	t	r	wh	r	m	n	wh	t	w	r	w
t	h	'	t	'	h	'	m	n	h	t	v	'	v
t	'	r	t	r	'	r	m	n	'	t	v	r	v
t	s	l	t	l	f	l	m	n	f	t	v	l	v

Which reveals the missing words:

	one	two	three	four	five	six	seven	eight	nine
Hawaiian	*kahi*	*lua*	***kolu***	*ha*	*lima*	*ono*	*hiku*	*walu*	***iwa***
Maori	*tahi*	*rua*	*toru*	*wha*	***rima***	*ono*	*whitu*	*waru*	*iwa*
Nuku Hiva	*tahi*	*'ua*	*to'u*	*ha*	*'ima*	*ono*	***hitu***	*va'u*	***iva***
Rarotongan	*ta'i*	***rua***	***toru***	*'a*	*rima*	*ono*	*'itu*	*varu*	*iva*
Samoan	*tasi*	*lua*	***tolu***	***fa***	*lima*	*ono*	*fitu*	***valu***	*iva*

29 OK COMPUTER

a)   10 + 9 = 19
     6 + 1 = 7
     5 x 5 = 25

b)   *tadang + miit = 26 = aleeb madi*
     ataling madi – aleeb = 22 = bokob madi

Let's number the equations like this:

[i] *asumano × aleeb = bokob*   [vi] *ataling × ataling = tadang madi*

[ii] *asumano × ataling = tadang*   [vii] *asumano + ataling = feet*

[iii] *bokob × ataling = ataling madi* [viii] *feet + miit = feet madi*

[iv] *bokob × asumano = nakal madi* [ix] *tadang + ataling = tadang madi*

[v] *asumano × feet = feet madi*

From [vi], *ataling* must be 2, 3, 4, or 5; otherwise *tadang madi* is more than 30, which is not allowed.

Let's say *ataling* = 2. If it is, then from [vi], *tadang madi* = 4. But from [ix] this would mean that *tadang* = 2. But we initially assumed that *ataling* is 2. Contradiction. So *ataling* is not 2.

Let's say *ataling* is 3. From [vi] and [ix] this makes *tadang* = 6. From [ii] this then makes *asumano* = 2, and from [vii] this makes *feet* = 5. If *aleeb* is 7 or greater, then from [i] *bokob* is 14 or greater, and thus from [iii] *ataling madi* is 42 or greater, which is not allowed. So *aleeb* must be less than 7. The only candidate left is *aleeb* = 4, which makes *bokob* = 8 and, from [v], *feet madi* = 10. But if this is so, from [viii] *miit* = 5, which is a contradiction, since *feet* = 5. So *ataling* is not 3.

Let's say *ataling* is 5. From [vi] and [ix] this makes *tadang* = 20, which from [ii] makes *asumano* = 4. So from [i], *bokob* is at least 8, which means via [iii] that *ataling madi* is at least 40, which is not allowed. So *ataling* is not 5.

By a process of elimination, we arrive at the conclusion that *ataling* = 4. This makes (from [vi]) *tadang madi* = 16, (from [ix]) *tadang* = 12, (from [ii]) *asumano* = 3, (from [vii]) *feet* = 7, (from [v]) *feet madi* – 21, and (from [viii]) *miit* = 14. *Aleeb* cannot be 5 or larger, since (from [i]) this would make *bokob* at least 15 and (from [iii]) *ataling madi* at least 60, which is not allowed. So *aleeb* must be 2, which makes *bokob* = 6, *ataling madi* = 24, and *nakal madi* = 18.

From the extra equations we can deduce that *awok* = 5, since if it was larger its square would be greater than 30, and the other numbers with squares lower than 30, which are 2, 3 and 4, are already taken care of. Thus *asumano madi* = 25, and lastly *maakob* = 1.

Let's list what we have:

1	*maakob*	6	*bokob*	18	*nakal madi*
2	*aleeb*	7	*feet*	21	*feet madi*
3	*asumano*	12	*tadang*	24	*ataling madi*
4	*ataling*	14	*miit*	25	*asumano madi*
5	*awok*	16	*tadang madi*		

Can you see the pattern? *Madi* must mean "the difference from 28," since, for example, 25 is 3 from 28, 24 is 4 from 28, and so on. So:

3	*asumano*	25	*asumano madi*
4	*ataling*	24	*ataling madi*
7	*feet*	21	*feet madi*
12	*tadang*	16	*tadang madi*

Thus *nakal* = 10, and *beeti* = 9, and the rest follows.

For the sake of completeness, here's how the Tifal body-tally system works. The numbers 1 to 14 correspond to 14 parts and sections of the body, starting with the little finger and ending with the nose. Counting back down from the nose after 14, the second counting of the left eye, which is 15, is equivalent to the difference from 28.

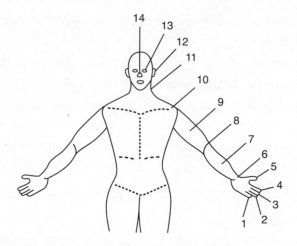

Source: *History of Number: Evidence from Papua New Guinea and Oceania*, by Kay Owens, Glen Lean, Patricia Paraide, and Charly Muke.

30 CELTIC COUNTING

The first thing to notice is that all the numbers have a different digit in the tens position. And in the units position only the 2 appears twice.

The only single digit is 9. The simplest word is *nuy*. A sensible hypothesis is therefore that *nuy* = 9.

In English, the word for 9 begins with an "n," as it does in many other languages spoken in nearby countries (such as French, German, and Spanish). One might expect the Manx word for 9 to also begin with an "n." Certainly, it's the obvious place to start.

The next simplest word is *daa-yeig*. The smallest remaining number is 12. How might *daa-yeig* mean 12? Maybe ten + two, or two + ten? The *yeig* part looks quite similar to *jeig*, which appears another four times in our list of number words. So let's hypothesize that *yeig/jeig* = 10, and that *daa* = 2. (You might feel an extra degree of security here, since in nearby Romance languages the word for "two" often begins with a "d," such as *deux* in French and *dos* in Spanish.)

I mentioned already that two of the given numbers end in the same digit, 2. However, two of the Manx terms end in *as feed*, and two end in *as daeed*. Clearly, then, the end of the Manx term does not correspond to the last digit of the numeral, as happens in English. Something else is going on.

Let's take a step back. All the Manx numbers in the list have a different tens digit, but in two separate instances they use the same terms: *as feed* and *as daeed*. Might there be any other language that uses the same terms to refer to different tens digits? *Oui!* In French, twenty is *vingt*, and eighty is *quatre-vingts*, or 4 × 20. Perhaps Manx is a little like French?

Let's order our numbers in batches of 20:

0–19	9, 12
20–39	25, 38
40–59	42, 57
60–79	66, 74
80–99	93

And let's look at those recurring terms in the Manx:

as feed	appears twice
as daeed	appears twice
tree feed	appears twice
kiare feed	appears once

Looks like we're getting somewhere. It's looking like *feed* = 20.

The word *feed* appears five times. OK, so *daeed* is not *feed*, but the words share the *-eed* ending, and—hang on—didn't we assume that *daa* = 2? So maybe over time *daa feed*, which would be 2 × 20, became *daeed*. Pronunciation changes with time, and letters and sounds can merge when spoken quickly. Numbers, which are commonly used in rapid speech, will be especially likely to develop irregularities. So *daeed* = 40 seems very plausible. Which would make *tree feed* = 60 and *kiare feed* = 80.

Carrying on down this route, we can take the remaining words two by two. *Hoght-jeig as feed* and *queig as feed* must correspond to 25 and 38. Since *feed* = 20, *hoght-jeig* and *queig* must correspond to 5 and 18. The simplest word, *queig*, is surely 5. And we had already supposed that *jeig* = 10. So *hoght* = 8, and *hoght-jeig* = 18.

Now to *jees as daeed* and *shiaght-jeig as daeed*. These numbers must correspond to 42 and 57. Using the same logic as above, *jees* = 2, *shiaght* = 7, and *shiaght-jeig* = 17.

Although we seem to have run into a contradiction. Didn't we say that *daa* was 2? Well, yes and no. We have only seen *daa* in compounds—in *daa-yeig*, and *daeed*. So maybe 2 has different forms when it's in a compound and when it's on its own. This wouldn't be so different from English, in which we use, for example, "three" when the number is on its own, but "thir-" when it's in compounds such as thirteen and thirty.

Now to *tree feed as kiare-jeig*, *tree feed as shey*, and *kiare feed as tree-jeig*, which must correspond to 66, 74, and 93.

We note that word order seems to change at *tree feed*, or 60, and that *tree* = 3, *kiare* = 4, and *shey* = 6.

So, the full solution is:

9	*nuy*
12	*daa-yeig*
25	*queig as feed*
38	*hoght-jeig as feed*
42	*jees as daeed*
57	*shiaght-jeig as daeed*

66   *tree feed as shey*
74   *tree feed as kiare-jeig*
93   *kiare feed as tree-jeig*

For the sake of completeness, the numbers from 1 to 10 in Manx are:

nane, daa, tree, kiare, queig, shey, shiaght, hoght, nuy, jeih

LINGO BINGO

Chinese Compound Words

1. a	4. d	7. a	10. b
2. c	5. d	8. c	11. b
3. b	6. a	9. c	12. a

4. Decipher Yourself!

31 GRIPPING REEDS

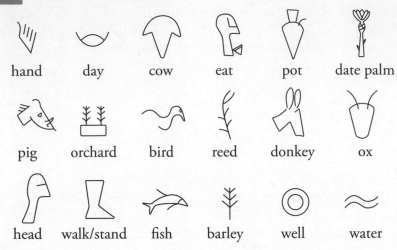

hand	day	cow	eat	pot	date palm
pig	orchard	bird	reed	donkey	ox
head	walk/stand	fish	barley	well	water

32 SHUT UP SON!

1. f	4. b	7. a	*sheru* =
2. c	5. e	8. d	*qula* =
3. g	6. h		

Each cuneiform symbol represents a syllable, such as *ba, ma,* or *la.*

First, we need to work out the direction of the writing. No two words share the same first symbol, but in some words the last symbol is the same. Likewise, no two pronunciations share the same first syllable, but some share the same last syllable. So we can deduce that the writing reads from left to right.

Our way in is with the symbol 𒇻, which appears at the end of three words, and at the beginning of one. The syllable *lu* also appears at the end of three words and at the beginning of one, so it would appear that 𒇻 = *lu*. One of the words ending in 𒇻 has three syllables (*nubalu*), so this is probably the word with three symbols. From that we can deduce that 𒉡 = *nu*, and 𒁀 = *ba*. Thus 𒇻 𒉡 must be *lanu,* and 𒁀 𒇻 = *balu.*

The remaining words with 𒈷 must be 𒈷 ✹ 𒀳 𒈠𒉿 𒈷 = *lushepisamma*, and 𒍝 𒈷 = *qulu*. That gives us 𒍝 = *qu*, so 𒊑 𒍝 = *ruqu*. Finally, 𒈠 𒊑 = *maru*, and the remaining word is 𒌑 𒈷, which is *ubla*.

33 THE KINGS OF OLD PERSIA

I've already done the heavy lifting, which reduced the texts to these two sentences:

> Text 1:   A-m B C B B-n D-p E
> Text 2:   F B C B B-n A-q B-p E

I also told you that these texts translate as:

> "X, the great king, the king of kings, the son of Y . . ."

In the royal title, "king" is the most common word, appearing three times. In Texts 1 and 2, the most common element is B. Let's assume, therefore, that B = "king," which gives us:

> Text 1:   A-m **king** C **king king**-n D-p E
> Text 2:   F **king** C **king king**-n A-q **king**-p E

It would appear likely that the phrase "great king, the king of kings" is thus "**king** C **king king**-n," since this phrase is independent of the names of the kings and is replicated in both texts. So C = "great," and the suffix *-n* is likely to indicate the genitive (belonging). (This is not unusual. In English we sometimes add the suffix *-'s* to indicate belonging. The word "king's," for example, means "of the king.")

We now get:

> Text 1:   A-m, **great king, king of kings**, D-p E
> Text 2:   F **great king, king of kings**, A-q **king**-p E

(The Old Persian is written as "king great," but in English we reverse the word order.)

The other word that appears in both texts is E. Looking at the Sassa-nid sentence, it would be fair to assume that E = "son." Thus:

Text 1:   A-m, **great king, king of kings,** D-p **son**
Text 2:   F **great king, king of kings,** A-q **king**-p **son**

The first text seems to be saying that A is the son of D.

The second text seems to be saying that F is the son of King A.

We can assume that D is not a king, since if he were the inscription would mention it. In other words, King A is the son of D, who is not a king. And King F is the son of King A.

If you look at the Achaemenid family tree on page 89, the only kings with fathers who weren't kings were Cyrus and Darius. So A must be Cyrus or Darius.

Let's assume A = Cyrus. Thus D = Cambyses (from Text 1) and F = Cambyses (from Text 2). But this is a contradiction, since D and F are different names.

So A = Darius, and thus D = Hystaspes and F = Xerxes.

The full transcription is therefore:

Text 1: **Darius, great king, king of kings, son of Hystaspes.**
Text 2: **Xerxes, great king, king of kings, son of King Darius.**

34 CHAMPAGNE FOR CHAMPOLLION

You may have been distracted by the images. In fact, proper names in Egyptian hieroglyphs are spelled phonetically. Each figure represents a letter of the Latin alphabet. I mentioned that transcribed cartouches should be read left to right. The Rosetta stone cartouche reads P-T-O-L-M-E-S. Four of these symbols appear in the Philae cartouche. I also told you that the ⬭ and the ⌂ have the same meaning, which from the Rosetta stone is deciphered as the letter T.

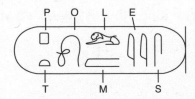

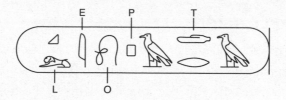

By filling in the Philae cartouche we get

*-L-E-O-P-?-T-!-?

where *, ?, and ! are unknowns. Note that ? appears twice. Think of all the famous Egyptian rulers you've ever heard of. The solution is: CLEOPATRA.

35 DEATH ON THE NILE

"Tutankhamun, ruler of Heliopolis of Upper Egypt"

In other words, the most famous ancient Egyptian ever (apart from Cleopatra, perhaps).

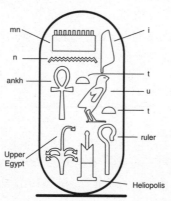

We know from the previous problem that the two hemispheres are "t"s. The only missing piece of information is that the single reed is an "i." You may have deduced it was some kind of vowel sound.

The direction it needs to be read is roughly midde-top-bottom. The top three symbols spell out *i-mn-n*, and the middle ones *t-u-t-ankh*. Put them together and you get:

t-u-t-ankh-i-mn-n

36 PURPLE REIGN

You know that the Phoenician alphabet only has consonants.

A scan for patterns reveals that the symbols 𐤁 and 𐤋 appear together in two words. Are any sequences of consonants in the English words repeated twice? Yes, "b-l" in Ri**bl**ah and E**bl**a. Let's assume that 𐤋 = "b" and 𐤁 = "l."

This would mean, however, that 𐤁𐤋𐤄 is of the form "l-b-?". There are no words of this form, so we are led to the realization that Phoenician is written from right to left.

Thus, 𐤋 = "l" and 𐤁 = "b." By piecing together the words that use these letters we get:

𐤋𐤁𐤀	Ebla
𐤀𐤋𐤁𐤓	Riblah
𐤁𐤋𐤄	Halab

The word beginning with 𐤁, or "b," must be Beritos. So,

𐤑𐤕𐤓𐤁	Beritos

We know from Riblah and Beritos that 𐤓 is an "r." So that gives us:

𐤈𐤓𐤑𐤕	Tsarephath

The other three cities are Megiduw, Qadesh, and Aynuk, which are (in some as-yet unknown order):

𐤊𐤓𐤂𐤌

𐤊𐤍𐤆𐤀

𐤔𐤃𐤒

None of those symbols is found in any of the other names. However, the 𐤃 is repeated, and the only repeated consonant is "d." Based on word length and position of each in the word, we can deduce that:

𐤊𐤓𐤂𐤌	Megiduw
𐤊𐤍𐤆𐤀	Aynuk
𐤔𐤃𐤒	Qadesh

The city which is still a regional capital today? Beritos—now better known as Beirut, of course!

You may also have noticed that several of the symbols bear some kind of resemblance to Latin letters, which may have helped you:

ϟ – A (the pronunciation is a glottal stop, rather than a vowel)

ᐃ – D

ϟ – K

ㄥ – L

ᔦ – M

Ϙ – Q

37 A CRETAN CRUNCHER

1. ⼞Ϙ†	n) *agros*	8. ⼋ⵕ⼌	j) *tripodes*
2. ⼏ㄥ	b) *dōra*	9. ⼽⊕⼆	e) *thygatēr*
3. Aⵎ	d) *epi*	10. A⼂⊕	c) *heneka*
4. ⊕Ϙ	l) *chalkon*	11. ⼈⼂⼌	m) *chrysos*
5. ‡⼆	i) *patēr*	12. ⼌ㄷ	g) *meta*
6. ⼌⼈	f) *meli*	13. ‡ㄥ	h) *para*
7. ‡ⵎϘ	k) *pharmakon*	14. Ϙⵜⵕ	a) *Knossos*

If we look for patterns, we spot three words that start with the same character:

‡⼆
‡ⵎϘ
‡ㄥ

Linear B is a syllabic script. Which of the Ancient Greek words begin with the same syllables or characters?

meli/meta
para/patēr
chalkon/chrysos

We seem to have a problem. There doesn't seem to be a set of three words that start with the same syllable. However, there is *pharmakon*, which starts with *pha*. In the text, I said that there is a Linear B figure that can be written in the Latin alphabet as *pa*, *pha*, or *ba*. Let's assume, therefore, that ‡ stands for *pa/pha*.

The *pa/pha* word with three syllables is *pharmakon*, so this means we have *pharmakon* = ‡ℳ𝑃.

The other syllables in *para* and *patēr* are *ra* and *tēr*, which must be ⧣ and ᒻ, but as yet we don't know which is which.

We also see that ⧣ appears in one other word, Φ⊕⧣, as does ᒻ, in ᚷᒻ. The syllable *ra* appears in *dōra*, and *ter* appears in *thygatēr*. Comparing words with the same number of syllables, we get Φ⊕⧣ = *thygatēr*, and ᚷᒻ = *dora*, so ‡ᒻ = *para*, and ‡⧣ = *patēr*.

This is all fitting together very nicely, so we can feel confident that our initial assumption about ‡ was likely correct.

Our syllabary as it stands is:

‡	*pa/pha*
⧣	*tēr*
ℳ	*ma* or *rma* (probably just *ma*)
𝑃	*kon*
ᒻ	*ra*
ᚷ	*do*
Φ	*thy*
⊕	*ga*

Looking at other words that contain any of these symbols, we have ⊕𝑃, which according to the syllabary would be *ga-kon*. However, we don't have a *gakon* in the Ancient Greek pronunciations. The closest word is *chalkon*, which would mean that ⊕ = *cha* and *ga*. This seems likely: The sounds *ga* and *cha* are similar. So, just as ‡ = *pa/pha*, ⊕ = *ga/cha*.

The remaining words we can partially fill in are:

ᚻ𝑃†	* - *kon* - *
𝑃Ⱬᚷ𝑃	*kon*- * - *
Ᏽᛦ⊕	* - * - *ga/cha*

The only word that begins with anything like *kon*- is *Knossos*, so it seems likely that:

𝑃Ⱬᚷ𝑃	*kon-no-sos*

No other words contain the syllable *ko* in them. However, the *ko* is a similar sound to *go*, so perhaps  = *ko(n)/go.*

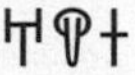 *a-go-ros*

No words end in *ga/cha*, but one ends in *ka*, which is a similar sound. Thus:

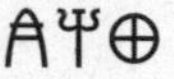 *he-ne-ka*

And:

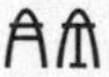 *(h)e-pi*

From *Knossos*, we know that  = *so(s)*, so:

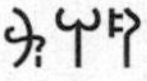 *chrysos*

The remaining words are *tripodes*, *meta*, and *meli*. Based on the number of syllables, we see that:

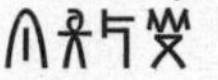 *tripodes*

What can  be? It must either be *ta* or *li*. If we break down *tripodes* to *t-ri-po-des*, it would seem likely that  = *li/ri*, so we have:

 *meli*

 *meta*

38 MASTERS AND SLAVES

1. e
2. c
3. d
4. h
5. a
6. b
7. g
8. f

All the phrases have the same structure. They describe an owner and a thing owned. For example, in the phrase "the donkey of the master," the master is the owner and the donkey the thing owned.

The Ancient Greek sentences have four words. The first is *ho/hoi* and the second is *tu/tōn*. It seems likely that these words will be a version of "the." The third and fourth words must be the nouns (i.e., "donkey," "master," "brother," "merchant," "son," "slave," or "house"), whose endings determine whether they are singular or plural, and whether they are the owners or the things owned.

The roots of the words are:

hyi-, *dul-*, *cyri-*, *oic-*, *on-*, *adelph-*, and *empor-*

The endings are:

-ōn, *-os*, *-oi*, and *-u*

We solve the problem using logical deduction and a process of elimination.

First, the word *cyri-* appears four times, and "master/s" appears four times, so our first deduction is that *cyri-* means "master."

If we look at the first nouns in each sentence, five end with *-ōn*, and three with *-u*. If we look at the second nouns, four end with *-os* and four with *-oi*.

Look at the English translations. The owners appear five times in plural form, and three times in singular form. The things owned appear four times in singular form, and four times in plural form. We can deduce, therefore, that *-ōn* refers to an owner in plural form, and *-u* is an owner in singular form.

Thus, *cyriōn* means "masters," and *cyriu* means "master," when they are owners. So:

h) *hoi tōn cyriōn hyioi*   4) the sons of the masters

e) *ho tu cyriu onos*   1) the donkey of the master

We can deduce that *on-* is "donkey," and *hyi-* is "son," so:

d) *hoi tōn onōn emporoi*   3) the merchants of the donkeys

a) *ho tōn hyiōn dulos*   5) the slave of the sons

From these sentences we learn that *empor-* means "merchant" and *dul-* means "slave." We now also know that things being owned have *-os* for singular and *-oi* for plural. And with that information we can deduce the rest.

Note that the order of the nouns in Ancient Greek is the reverse of that in English. It also might have helped that some of the Ancient Greek words look like words with similar meanings in English: *empor-* is like "emporium," a market; *dul-* is like "doula," a female helper/nanny; and full marks if you knew that "Philadelphia" means "brotherly love," from *adelph-*, "brother," and *phil-*, "love."

39 AND THE OSCAR FOR OSCAN GOES TO ...

When I told you that you might need time to reflect, I meant it literally!

Look at the Cippus Abellanus inscription when reflected in a mirror, as shown below. Immeditaely, you can see words that look like NUST, NUS, EISEI, TEPEI, which are the same as or similar to *púst, pús, eisei,* and *terei,* which appear in the list of possible words. We can deduce that the text runs right to left, and that many of the letters are mirror images of letters from the Greek and Roman alphabets. (I did tell you in the buildup that Oscan helps to fill in the gaps in how the Latin alphabet evolved from the Greek one.)

You might have come to the realization about word direction by seeing that on the top line of the inscription two Oscan words are identical, except that one has an extra "T"; likewise, two of the words in Latin script, *pús* and *púst,* differ only by a "t." Or maybe the backward "E"s suggested *eisei* and *terei.* Anyway, once this realization is made, you can start to deduce the pronunciation of letters that bear little relationship to Latin or Greek. For example, ⟨ is an "f." Vowels are sometimes joined to their following or preceding consonants.

eisei	fufans	**feihúis**	**amfret**
pússtis	**terei**	svai	ehtrad
pidum	**fisnam**	**pús**	inim
púst	anter	prúftú	eisúd

The remaining words are *nep Abellanus nep Nuclanus*. (We decipher the "n" from seeing how the "n" in *fisnam* attaches to the vowel, we get the "p" from *pús*, and so on.) The translation includes the phrase "neither the inhabitants of Abella nor the inhabitants of Nola," so it seems likely that:

a) *nep* is neither/nor.

b) *Abellanus* is the inhabitants of Abella.

c) *Nuclanus* is the inhabitants of Nola. (In fact, it is Nuvlanus, since the reversed "C" is actually a "v.")

40 NORSE CODE

1. ᛒᚠᛚᛗᚱ	a) Baldur	7. ᛋᛟᚱᚦ	d) Earth
2. ᚦᛟᚱ	j) Thor	8. ᛗᛗᛏᛁᛟᛗᚱ	b) Dallinger
3. ᛁᚦᚢᛏᛏ	g) Ithun	9. ᚠᚱᛗᛁᚱ	f) Freyr
4. ᛗᚠᚷᚱ	c) Day	10. ᛟᛗᛁᛏ	k) Odin
5. ᛏᛟᛏᛏ	h) Night	11. ᛋᛟᛚ	i) Sun
6. ᚠᚱᛗᛁᛋᚠ	e) Freya		

First let's check the direction of the text by looking at the letters at the beginning and end of the words.

Two runic words begin with ᛗ, and two begin with ᚠ. Five runic words end with ᚱ, and two with ᛏ.

Two English words begin with "D," and two with "F"; three end in "r," and two in "n."

Even accounting for the two "missing" English words (which we need to deduce), it seems clear that the reading direction is left to right. You might also use your visual intuition: The ᚱ looks very like an "R." And you know that runes and Latin script have a common source.

So let's assume a reading direction of left to right. That would mean that ᛗ and ᚹ are "D" and "F," although as yet we're not sure which is which. Looking at the length of the words, it's likely that:

ᛗᛖᛚᛁᛜᛖᚱ Dallinger

ᛗᚨᚷᚱ Day

And thus:

ᚹᚱᛖᛁᛜᚠ Freya

ᚹᚱᛖᛁᚱ Freyr

(Because it's likely that ᚱ = "r," from Dallinger and from the familiar shape. The runic word for "day" has an ᚱ. But maybe this is because the word "day" is not the transliteration of a proper name, unlike the names of the gods.)

The ᛗ/D is also present in ᛒᚨᛚᛗᚱ, which must be Baldur.

There are seven runes in Dallinger. By crosschecking with Baldur, we see that ᛚ must be "l," so the rune ᛁ is probably "i." Which makes:

ᛁᚦᚢᚾᚾ = Ithun

The double ᚾ is probably just "n-n." You know (since I told you) that "th" is a single letter, thus ᚦ = "th." So the remaining words now look like this:

2. ᚦᚩᚱ   th-*-r

5. ᚾᚩᛏᛏ   n-*-*

7. ᛋᚩᚱᚦ   *-*-r-th

10. ᚩᛗᛁᚾ   *-d-i-n

11. ᛊᚩᛚ   *-*-l

7 is clearly "earth," and we can see that 2 and 10 must be Thor and Odin, the missing gods (and the most famous ones), leaving 5 and 11 to be "night"/"day." The most probable candidate for "night" is the word beginning with "n," which ends in two letters that are similar to a Latin "t," which makes 11 "sun."

LINGO BINGO

Portmanteau Words

1. a) bro + Brazilian
2. c) calf + ankle
3. d) spoon + fork
4. c) flugelhorn + trumpet
5. a) interrogative + bang (The interrobang is the ‽, the superimposition of a question mark on an exclamation mark.)
6. d) three + couple
7. a) snow + ice
8. c) croissant + donut
9. b) jeans + shorts
10. c) mist + drizzle

5. Relative Values

41 ■ THE FARFAR NORTH

mormor: maternal grandmother (mother's mother)
morfar: maternal grandfather (mother's father)
farmor: paternal grandmother (father's mother)
farfar: paternal grandfather (father's father)

42 ■ MY ROMAN FAMILY

Camilla is Cato's *amitae adneptis*
Rufus is Octavius's *avunculi trinepos*
Your daughter is Vita's *materterae neptis*
Your great-grandson is Julia's *patrui abnepos*

To work this out, start by describing in English the Latin terms in the question:

Titus is the great-great-grandson of your uncle.
Septimus is the great-grandson of your aunt.
Livia is the granddaughter of your aunt.
Florentina is the great-great-great-granddaughter of your uncle.
Flavia is the great-great-great-great-granddaughter of your aunt.

The second word in each Latin phrase contains the root *nep*. It would seem likely that the root means "grandchild" and the completed word refers to some generation of grandchild. By counting the generations, we deduce that the word for "grandson" is *nepos* and "granddaughter" is *neptis*. "Great-grandson"/"daughter" is *pronepos/neptis*, and the next generations go *abnepos/abneptis*, *adnepos/adneptis*, *trinepos/neptis*.

The first word refers to the uncle/aunt. It's a short step to realize that Latin distinguishes between maternal and paternal aunts (*matertera*, *amita*) and maternal and paternal uncles (*avunculus*, *patruus*). The *-ae* and *-i* endings denote the genitive forms.

43 MEET THE RELATIVES

a) Friulian b) Faroese c) Breton d) Hindi
e) Upper Sorbian f) Limburgish

General knowledge and information picked up from other parts of this book will be helpful here. For example, English, Faroese, and Limburgish are Germanic languages, so we can assume that these will be the most similar. We know from chapter 2 that the letter ð exists in Scandinavia, so b) is a likely contender for Faroese. Likewise, *ich* is "I" in German, so f) is a good candidate for Limburgush. The language with words that are least like the others is d), so this is probably Hindi. Language a) seems to have similarities to Romance languages like French and Italian, so it's fair to guess that this is Friulian, which is spoken in Italy.

When it comes to deciding the Celtic and the Slavic, we're left with c) and e). We've encountered Celtic languages such as Welsh, Irish, and Manx in this book already, and they have more in common with c). So by a process of elimination c) is Breton, and e), with its Slavic č, is Upper Sorbian.

44 RICE WITH THE GRANDKIDS

I asked you to find the English phrase that does not include the word "grandchild," but which has a meaning that relates to being a descendent of someone or something. The answer must be "shoot of rice." So, we know that one of the following words is "shoot of rice," or, literally, "*grandson-rice*":

zafim-bary   zafin-dohalika
zafim-paladia   zafin-kitrokely

The word "rice" is one of *bary, dohalika, paladia,* or *kitrokely.* In the English expressions the word "rice" appears again in "rice field." So presumably the Malagasy term for "rice field" contains one of the words *bary, dohalika, paladia,* or *kitrokely* combined with another single word. The only Malagasy word that fits is *tanim-bary,* so we can assume that *tanim* = "field," *bary* = "rice," and *zafim-bary* is "shoot of rice." *Tanim* does not appear in any of the other Malagasy words, nor does "field" appear in any of the other English meanings. This gives us extra confidence that we're on the right track.

Of the remaining words, *mahambozona* and *mahandohalika* seem related, because they have similar prefixes. Likewise, there are two English expressions beginning "one who." It's likely, therefore, that *mahambozona* and *mahandohalika* refer to the person who can carry something on their neck, and the one who can get on their knees, although we don't yet know which is which.

Mahandohalika also seems related to *lohalika*. *Mahambozona* does not resemble any other word. In the English expressions, the word "knee" appears twice, but "neck" appears only once. We can thus hypothesize that *lohalika* is "knee," *mahandohalika* is the person who can get on their knees, and *mahambozona* is the person who can carry something on their neck.

By a process of elimination we're left with *kitrokely* and *hafaladia*, which must be "ankle" and "up to the sole," although as yet we don't know which is which.

Now back to the "grandchildren" terms. We know that when *lohalika* is compounded it becomes *-dohalika*. So *zafim-dohalika* has the literal meaning "grandson-knee." The other two *zafim/n* words must be literally "grandson-ankle" and "grandson-sole." But which is which? *Kitrokely* becomes *zafin-kitrokely*. But *hafaladia* becomes *zafim-paladia*. The word *hafaladia* seems to contain an extra prefix *ha-*, which presumably means "up to the." So *kitrokely* is "ankle" and *hafaladia* is "up to the sole." Thus *zafim-paladia* is "grandson-sole," and *zafin-kitrokely* is "grandson-ankle."

In English we have the terms for three generations of grandchildren, and in Malagasy we have "grandson-knee," "grandson-ankle," and "grandson-sole." It is natural to assume that as the generations get more distant, the body parts that correspond to them get further down the body. So the correct matches are:

hafaladia	up to the sole
kitrokely	ankle
lohalika	knee
mahambozona	one who can carry something on his neck
mahandohalika	one who can get on his knees
tanim-bary	rice field
zafim-bary	shoot of rice (departing from the stem)
zafin-dohalika	great-great-grandchild
zafim-paladia	great-great-great-great-grandchild
zafin-kitrokely	great-great-great-grandchild
zafy	grandchild

For the sake of completion, "great-grandchild" in Malagasy does not have a corresponding body-word to indicate the generation. It is *zafiafy*, a partial reduplication of the word *zafy*.

45 THE COUSIN WHO HUNTS DUCKS

Here are two possible sentences for each phrase:

1. The cousin who hunts ducks is a good hunter.
   The cousin who hunts ducks out on weekends.

2. The florist sent the flowers to his lover.
   The florist sent the flowers was surprised to receive such an unoriginal gift.

3. The cotton clothing dries faster.
   The cotton clothing is made of is imported.

4. The woman who whistles tunes is tone-deaf.
   The woman who whistles tunes pianos.

5. We painted the wall with blue paint.
   We painted the wall with cracks.

6. I convinced her children to go to sleep.
   I convinced her children are evil.

7. When the baby eats food the mom is overjoyed.
   When the baby eats food gets thrown.

8. Mary gave the child the dog for Christmas.
   Mary gave the child the dog bit a plaster.

9. The girl told the story quietly.
   The girl told the story cried.

10. That John is never here, the lazy so-and-so.
    That John is never here hurts.

46 AMY, SUE, AND BOB, TOO

This problem reveals that verb endings reflect whether or not the subjects of the verb are in sibling relationships.

If the two subjects are siblings, the verb form is *puddeniyithnu*.

If the two subjects are not siblings, the verb includes the suffix

-nintha if they are two males, or

-ngintha if they are two females, or a male and a female.

If there are three subjects, two of whom are siblings, the verb takes the suffix

-neme if all are males

-ngime if at least one is female.

From which we can deduce that:

1. *Bob i Jim puddeniyithnunintha*
2. *Des i Daisy puddeniyithnungintha*
3. *Bob, Jim, i Dave puddeniyithnuneme*
4. *Fred i Dave puddeniyithnu*
5. *Jane i Dave puddeniyithnu*
6. *Des, Jane, i Fred puddeniyithnungime*
7. *Daisy, Jane, i Sue puddeniyithnungime*

We can also see that if there are three subjects, all of whom are brothers, the stem changes to:

puddiniyithnu (in which the first "e" changes to an "i").

In total, there are more than 40 possible forms for this verb, depending on the number of people who are teaching each other and their sibling relationships.

47 MY WIFE'S FATHER'S MOTHER'S BROTHER

The male/female names for the skins are:

1. Jakamarra/Nakamarra
2. Jampijinpa/Nampijinpa
3. Japanangka/Napanangka
4. Jungarrayi/Nungarrayi

5. Japaljarri/Napaljarri
6. Japangardi/Napangardi
7. Jupurrula/Napurrula
8. Jangala/Nangala

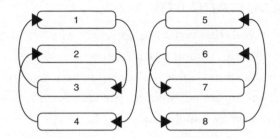

Here are the rules again: Horizontal rows indicate marriages, and arrows point from mother to child. I'm going to define a shorthand for some simple relationships:

Husband to wife (H2W) move across to the other side of the row

Wife to husband (W2H) move across to the other side of the row

Mother to child (M2C) move along the arrow

Child to mother (C2M) move backward along the arrow

Father to child (F2C) move across to the other side of the row, and then along the arrow (H2W + M2C)

Child to father (C2F) move backward along the arrow and then across the row (C2M + W2H)

Sibling to sibling (SIB) all siblings have the same skin, so no move

Okay, we're now ready. We are told Nakamarra is skin 1, so let's start with sentence (ii), and break it down. It is saying: If I am 1, and if you apply the rule for sibling, and then the rule for father to child, you get Jupurrula. In our shorthand:

1 → SIB → F2C = Jupurrula.

(The → sign shows the direction of our deduction.)

When we follow the diagram, this gives us:

1 → 1 → 7 = Jupurrula. Thus, 7 is Jupurrula for men, and Napurrula for women. Knowing that Napurrula is 7, let's go to (iv).

Napangardi → W2H → SIB → W2H → C2F → C2F → C2M = 7.

But we can go the other way, too:

7 → M2C → F2C → F2C → H2W → SIB → H2W = Napangardi.

7 → 6 → 3 → 6 → 2 → 2 → 6 = Napangardi. So Napangardi is skin 6.

Take a breath, and look at the daunting statement in (vi). If you look closely it is repeating the phrase "wife's father's mother's brother" several times. This is the relation:

H2W → C2F → C2M → SIB, which simplifies as H2W → C2M → W2H → C2M, which you may notice always takes you back to where you started. We can thus ignore all appearances of "wife's father's mother's brother" and are left with

Japanangka → H2W = 7. So Japanangka/Napanangka is 3.

Now to statement (iii). The difficulty here is that we have two types of grandfather (maternal and paternal), and we don't know which is Jungarrayi and which is Jupurrula (7). So, of the following pairs of statements, either A or B is true:

A) Nampijinpa → C2M → C2M → C2F = 7 (maternal).

Nampijinpa → C2M → C2F → C2F = Jungarrayi (paternal).

B) Nampijinpa → C2M → C2M → C2F = Jungarrayi (maternal).

Nampijinpa → C2M → C2F → C2F = 7 (paternal).

By working out the possibilities we see that if 7 is the (mother's) paternal grandfather, (i.e., pair B is correct) then Nampijinpa is 6, which is already taken. So pair A is correct: 7 must be the (mother's) maternal grandfather, which means that Jampijinpa/Nampijinpa is 2, and Jungarrayi/Nungarrayi is 4.

It's all coming together. From (i) Jangala → F2C → 2, we deduce that Jangala/Nangala must be 8, leaving Japaljarri/Napaljarri as 5.

48 BURMESE BABIES

The traditional Burmese way to name children is to use the day of the week on which they were born to determine the first letter. If you listed the names in the question according to the days of the week, you would find that the first letters are almost all the same, and where there is a difference the sounds are similar.

day	boy	girl	first letter/s
Mon	kauŋ myaʔ khaiŋ miŋ thuŋ	kethi thuŋ	k
Tues	zeiya cɔ shaŋ thuŋ	su myaʔ so susu wiŋ shu maŋ cɔ	z, s, sh
Weds	lwiŋ koko wiŋ cɔ auŋ	wiŋ I muŋ yiŋyiŋ myiŋ yadana u	w, lw, y
Thurs	phouŋ naiŋ thuŋ pyesouŋ auŋ myo khiŋ wiŋ	mimi khaiŋ paŋ we	p, m
Fri			
Sat	thɛʔ auŋ tiŋ mauŋ laʔ	thouŋ uŋ tiŋ za cɔ	t, th
Sun			

Once you work out the days of the week for February 3, 4, 15, 16, 26, and 27, and look for similar first letters, you get:

khiŋ le nwɛ	Monday, February 3
so mo cɔ	Tuesday, February 4
daliya	Saturday, February 15—"d" is phonetically similar to "n" and "t"
e tin	Sunday, February 16—no names in the table begin with an "e," nor are there any Sunday birthdays, so by default we deduce that Sunday names begin with "e"

ye auŋ naiŋ   Wednesday, February 26
phyuphyu wiŋ   Thursday, February 27

The system is based on the Burmese eight-day astrological week, in which Wednesday is divided into morning and afternoon. (Morning babies are "w" and afternoon are "y.") Each of the eight days is associated with a different compass direction.

49 A TREE IN ICELAND

There is one foreign surname among the guests, so let's isolate this branch of the tree:

Ingimundur Sigurðarson Bergmann
Jón Oddsson Bergmann
Rakel Ragnheiðardóttir Bergmann
Róbert Bergmann Gunnarsson
Sigurður Jónsson Bergmann

We can make a partial family tree, since Ingimundur is the son of Sigurður, who is the son of Jón. Rakel is the daughter of Ragnheiður, who we know is female because she is the daughter of Jakob. We don't know who she is married to quite yet. Nor can we say anything about Róbert.

The three children of Guðrún and Jakob must either be Guðrúnarson/dóttir or Jakobsson/dóttir. The candidates are Daníel Guðrúnarson, Ragnheiður Jakobsdóttir, Sara Jakobsson Þorarinssonar, and Steinunn Jakobsdóttir. However, we know that Jakob's father is Kristján, so we can reject Sara, since if she were Jakob's daughter she would be Sara Jakobsson Kristjánssonar. Thus Sara is the daughter of Jakob Þorarinsson.

Margrét is the daughter of Steinunn (and granddaughter of Jakob). Daníel Steinunnarson Guðrúnardóttur is also a son of Steinunn and thus a brother of Margrét. Who might Jakob Þorarinsson be married to? We can eliminate Margrét, since she is only 11 years old. The only other option is Steinnun, since, as we will see, Jakob cannot be married to Ragnheiður, nor to any of her descendants.

Ragnheiður's daughter is Rakel, who is a Bergmann, so Ragnheiður must be married to the oldest generation of the Bergmanns—i.e., Jón. The only other unused pieces in the jigsaw are Gunnar Gunnarsson, Stefán Gunnarsbur Gunnarssonar, and Róbert Bergmann Gunnarsson. Stefán is the nonbinary child of Gunnar. The fact that Róbert is both a Bergmann and the son of a Gunnar places him on the Bergmann branch, and the only way that can fit is if Rakel is married to Gunnar.

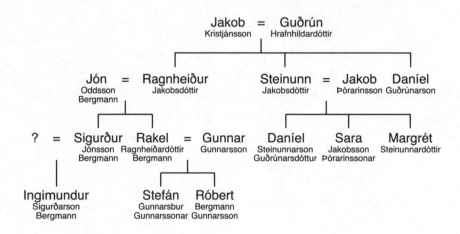

50 THE FAMILY TABLE

The knowledge that Alexa is the 1 of Edward, and Edward is the 3 of Alexa does not give us enough information to progress. There could be many possibilities for 1 and 3: mother/son, father/daughter, aunt/nephew, and so on.

However, Edward is the 4 of Iggy, and Iggy is the 4 of Edward. The relationship works both ways. This symmetry is the starting clue we need to solve the problem.

Of the terms in the list, the only possible fit for 4 is either "brother," "sister," or "cousin." We can eliminate "sister," since Edward is male. So we're left with "brother" or "cousin." "Brother" is the simplest solution, so let's start with that.

If Edward and Iggy are brothers, what do we know about Ollie? We know that Edward and Iggy are brothers to Ollie, but Ollie is not a brother to Edward and Iggy. Conclusion: Ollie must be female and she is the boys' sister. So 8 is "sister."

We can start to build the family tree:

Edward—Iggy (male)—Ollie (female)

Alexa has the same relationship (1) to Edward, Iggy, and Ollie. So 1 is either "mother," "grandmother," "aunt," or "cousin." It can't be "cousin," because then the siblings would have the same relationship to Alexa, and they don't. (Edward and Iggy are 3's to her, and Ollie is a 7 to her.) Nor can 1 be "grandmother," because each of the siblings would then all be "grandchildren" to her, and we know they have different relationships. ("Grandson"/"granddaughter" are not in the list of available words.) So 1 is either "mother" or "aunt."

If Alexa is Edward's, Iggy's, and Ollie's aunt, then 3 is "nephew," which means that Uzzie is Edward's nephew. And therefore Alexa is the "aunt of the uncle" of Uzzie, or his great-aunt, a relation that is not in the list. So Alexa is the mother of Edward, and Uzzie is Edward's son. Thus 1 = "mother," 3 = "son," and 7 = "daughter." We can now draw the family tree:

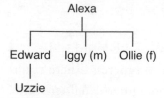

And thus the table of correspondences is:

1	mother
2	grandmother
3	son
4	brother
5	father
6	uncle
7	daughter
8	sister
9	aunt
10	grandchild
11	nephew

Before we finish we need to double-check that 4 is not "cousin." Let's assume that it is—i.e., that Edward and Iggy are cousins. They must also be cousins of Ollie. Yet the table states that Ollie is *not* a cousin to Edward or Iggy, which would be impossible. So 4 cannot be "cousin." Phew, we're done!

LINGO BINGO

Animal Sounds

1. b	4. c	7. d	10. c
2. a	5. a	8. c	
3. d	6. b	9. a	

6. Aiding and Alphabetting

51 THE WRITE WAY TO SPEAK

a) peaks   c) tap   e) back   g) dog

b) boot   d) cogs   f) piece/peace   h) peas

The "p" sound appears at the beginning of "peaks" and at the end of "tap." The "t" sound appears at the beginning of "tap" and at the end of "boot." So "peaks," "tap," and "boot" must be some combination of a), b), and c), because the first and last letters of d) are not in the first or last position of any other word. We can also hear that "peaks" and "cogs" are made up of four phonemes—"p-ee-k-s" and "c-o-g-z," so a) is "peaks," and the rest fall into place.

The only letter you need to deduce in the remaining words is the "d" in "dog." This can be guessed by noticing that the symbol is the same as the symbol for "t" but with the extra line segment. So the sound will be very similar to "t." In fact, we can see that an extra line segment is the difference between "s" and "z"; in other words, that line segment indicates that you should make the "voiced" version of that particular consonant sound, which in the case of "t" is "d."

52 FAST TALKING, FAST TYPING

1. B, E, C, A, D

2. The defendant: "Absolutely one hundred percent not guilty."

We'll piece this together using three observations that reinforce one another:

i) There are three lines in which the same eight consonants on the right of the asterisk have been pressed. In two of these instances all the consonants to the left of the asterisk have also been pressed. It would seem unlikely that in a system designed for speed and efficiency that these chords denote syllables. More likely, these lines are "dividers" that denote a change of speaker. We can see that the court speaks twice, and the defendant once, so this means that the two lines in which almost all consonants are pressed are likely to represent the court. Handily, one of the five pieces (B) begins with one of these lines, so we can tentatively say that piece (B) is the first one.

ii) Rewrite the dialogue in syllables. Written phonetically we get:
Court: ar / yu / re / di / to / en / ter / a / plee / at / this / taim
Defendant: yes / your / o / nuh /
Court: how / do / yu / pleed / to / cownts / wun / and / two

In the explanation of how stenography works, I told you that when "n" is at the end of a syllable, it is written as P B, and that "a" is A.

There are four places in which an "n" sound is part of the final consonant of a syllable:

en
cownts
wun
and

The syllables "cownts," "wun," and "and" appear consecutively. Piece (D) has three consecutive lines containing P B, so it's looking like piece (D) is where these syllables appear. If we look more closely, the syllable after "and" is "two," which fits. Since this is the end of the dialogue, we can be fairly sure that the final piece of text is piece (D).

There is one line in piece (E) that consists of a solitary letter A. This must be the only "a" in the text. The first line of piece (E) has the letters E and PB (which we know is "n"), so this line might be the syllable "en." The next line has TER, which could be "ter" and the final line has AT, which could be "at." It's looking likely that piece (E) contains the phrase "enter a plea at," which would mean that piece (E) is the second in the sequence.

iii) To reinforce our view that that piece (B) is at the beginning, the second line R U looks very much like an abbreviation for "are you."

Assuming that (B) is the first piece, (E) the second piece and (D) the final piece, we can begin to work out chords for consonants and vowels. For example, the penultimate line of (E) is PHRAOE, which we know must be the syllable "plee." Thus PHR is "pl" and AOE is "ee." The bottom line of piece (A) has PHRAOED, which must be "pleed," and we can feel very confident that piece (A) is the penultimate one in the sequence.

The other chords for consonants and vowels that appear in the problem (and which can be worked out now we know the order of the pieces) are:

Initial consonants

h	H
r	R
d	TK
t	T
p	P
l	HR
y	KWR
w	W
c(k)	K

Final consonants

n	PB
r	R
t	T
m	PL
d	D
s	S

Vowels

a as in "had"	A
e as in "bed"	E
o as in "hot"	O
u as in "bun"	U
ee	AO E
i as in "time"	AO EU
"ow" as in "how"	O U

Using this set, you can deduce that the final text is pronounced, syllable by syllable:

sl[vowel]t
lee
wun
hun
pers
[consonant]ot
[consonant][vowel][consonant, probably l]t
tee

The context makes it clear which words are abbreviations and which are syllables, and with some educated guesswork you will get the full answer. Maybe try slurring . . .

53 ■ GREAT MEMORIES

a) 3,276   b) 7,117   c) 745,821   d) 64,101,491,094

The system ascribes consonant sounds to each digit. What's initially confusing is that some digits have more than one sound, and the sounds "h" and "w" (and "y") are ignored.

0	s, (z)	5	l
1	t, d, (th)	6	sh, ch, tch, (j)
2	n	7	c, g
3	m	8	f, v
4	r	9	p, b

The sounds in parentheses are not used in the puzzle. The sound "z" needs to be deduced for the answers. Remember, it is the sounds that are transcribed, not the spellings. Thus the "ph" in "elephant" is not a "p" and an "h," but an "f." Once the numbers have been converted into consonants, vowels can be inserted anywhere to make words.

54 A GODLY SILENCE

A.	chocolate milk	(black-milk)
B.	Italy	(E-tall-E-country)
C.	drink	
D.	Iceland	(hard-water-country)
E.	milk	
F.	Benedictine	(black-monk)
G.	Christmas	(baby-God-day)
H.	toilet	(shame-house)
I.	snow	(white-rain)
J.	the Blessed Sacrament	(God-bread)
K.	dormitory	(sleep-house)
L.	England	(drink-T-country)
M.	barn	(cow-house)
N.	ice	(hard-water)
O.	cake	(sweet-bread)
P.	Cistercian	(white-monk)

Here's one way to solve the problem. The 12-sign appears as the second sign in three words: next to one that looks like a bull/cow, next to one that looks like a person sleeping, and next to one that looks like a person covering their eyes. The only animal-related word is "barn" and the only sleeping related word is "dormitory," so it would seem to make sense that the 12-sign means something like "shelter" (it means "house"), and that M is "barn" and K is "dormitory." The only other place of shelter is "toilet," and it makes sense that covering one's eyes is related to a toilet: If someone is on the john, one does not look at them!

The 1-sign is of a person drinking, so the likely meaning of D is "drink."

The only other sign that appears three times is the 3-sign, of a monk pointing at the floor. There are three countries in the list, so it could be that the 3-sign denotes "country," which would make sense, since pointing at the floor could mean something like "land." (In fact, it means "courtyard," which is similar.)

The three countries are Italy, England, and Iceland. I mentioned that two of the signs represent letters. A good candidate would be the 2-sign, which is a T, which would make this country Drink-T(i.e., tea)-Land. The 16-sign is the middle finger, possibly "i," the third of five vowels? We can hypothesize that L is England and B is Italy, which is "i"—tall— "i"-land. (You could have deduced this without knowing that the second sign is "tall," which is used phonetically, like a rebus.) So D is Iceland.

The first two signs of Iceland appear on their own as N. Thus N must be "ice." In fact, you could have hypothesized that Iceland was D in the first place by seeing that N is contained in D, and comparing this to the observation that "ice" is contained in "Iceland."

The remaining signs are now:

A, E, F, G, I, J, O, and P.

I told you that there are signs for "God," "black," and "white," each of which is used twice.

Christmas is the birthday of baby Jesus, the son of God. The 4-sign in G is of a monk cradling a baby, so G is a good candidate for "Christmas." One of the other signs in G is likely to be "God." In fact, only one of the other signs in G, the 19-sign, is used twice, so the 19-sign must be "God." The most likely remaining phrase that contains the idea of God is "the Blessed Sacrament." So J, which contains the 19-sign, must be "the Blessed Sacrament." How might you describe the Blessed Sacrament in two simple concepts, one of which is God? "God-bread," perhaps? Thus, the 10-sign might be "bread," which means that O is probably something edible. The only other edible item on the list is "cake." So O is "cake."

You might be wondering which of the English words contain the ideas of "black" and "white," which each appear twice. Well, snow is white, and chocolate can be black. If you read the preamble to the question, you will also know that the Benedictines are the Black Monks and the Cistercians are the White Monks.

The word we haven't mentioned is "milk." Look at E: Is this not the milking of an udder? Yes! A must be "chocolate milk," from which we deduce that the 5-sign is "black," and the other words fall into place.

55 PIZZAS AND VERMOUTH

By counting the Braille characters and the letters, you would have seen that there are more characters than letters. Some of the extra characters are for punctuation:

Precedes a capital letter	Precedes a number	Comma	Full stop	Exclamation mark	Question mark
⠠	⠼	⠂	⠲	⠖	⠦

Your alphabet table should have looked like this:

a	b	c	d	e	f	g	h	i	j
k	l	m	n	o	p	q	r	s	t
u	v	x	y	z					w

The most noticeable pattern here concerns the bottom two dots in each character. The letters on the first row never use the bottom two dots, the letters on the second row use only the bottom-left dot, and the letters on the third row always use both dots. (Except the "w" for the reason stated in the question.) The hardest part of this problem is to realize that, if we only consider the top four dots in each Braille character, then the characters in each column are identical. In other words, given any letter in the first row, if we fill in the bottom-left dot we get the letter underneath it in the grid, and if we fill in both dots on the bottom line we get the letter two rows down. The full table is therefore:

a	b	c	d	e	f	g	h	i	j
k	l	m	n	o	p	q	r	s	t
u	v	x	y	z					w

Finally, the digits 1, 2, 3, 4, 5, 6, 7, 8, 9, and 0 are the letters from "a" to "j" when preceded by the "number" pattern illustrated above.

The answer is thus:

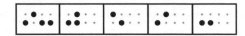

56 I • NEW YORK

a) Kathy	e) Jack	i) Heather
b) Elena	f) Gerald	j) Barb
c) Ivan	g) Lisa	k) Ashley
d) Carl	h) Fred	l) Dave

There are many ways to get the answer. In the preamble to the question I told you that capital letters are on four columns, which is why the initial letters of every name have four columns. I also said that the letters "e" and "t" consist of single dots, and that "a," "o," "i," "n," and "s" consist of two dots over two columns.

The name l) ends in a letter which is a single dot. Since no names end in a "t," this single dot must be an "e," and l) is Dave. This tells us the pattern for "a" and "v," which means we can deduce that c) must be Ivan. We then learn the pattern for "n," so b) is Elena, and the "l" gives us d), which is Carl. Then j) must be Barb, h) must be Fred, f) must be Gerald, i) must be Heather, and g) must be Lisa. We can deduce that there must be a single pattern for "sh" in Ashley—which has to be k)—and thus a) must be Kathy (from the "y"), which reveals a single pattern for "th." The remaining name, Jack, is therefore e).

We've done all of this deduction without looking at the initial letters, which are variations of the lower-case letters extended to cover four columns. In fact, to turn a letter into its capital you place the letter in the left-most column(s), and then add a row of dots (or a single dot) in the remaining columns. If the right-most column of the original letter has one dot in the top position, the extra dot(s) is/are in the bottom position. If the right-most column of the original letter has one dot in the bottom position, or in both positions, the extra dot(s) is/are in the top position.

Now to Orson. We know "r," "s," and "n" from Barb, Lisa, and Ivan. But we need to deduce the pattern for "o." We know that "o" has two dots over one or two columns. And we also know that:

a   i   n   s

There is only one possible remaining pattern (we can eliminate the pattern of two dots in the right column since this would be indistinguishable by touch from the "i"):

o

57   DOTTY ABOUT JAPAN

The given word is divided into four syllables.

ka   ra   o   ke

The *ka* and the *ke* are repeated in the problem, so we can deduce that:

c)   *sa-ke*

d)   *ka-ta-na*

The syllable *i* is the only other repeated syllable, so we can deduce that:

a) ⠿ ha-i-ku

b) ⠿ ko-i

Which leaves us with *a-ta-ri* and *ki-mo-no*, which must be these two—although as yet we don't know which is which:

e) ⠿

f) ⠿

The *-a* syllables all share a single dot in the top-left corner. The *-i* syllables all share the two top dots in the left column. This observation might lead you to realize that vowels are represented by the three dots in the top left, and the consonants by the three dots in the bottom right. Since Japanese Braille uses only five vowel sounds and nine consonant sounds, every syllable can be written with a total of 14 patterns, which is less than the 26 required for English Braille.

a i u e o

So d) is *ki-mo-no* and f) is *a-ta-ri*.

Finally, g) is *ka-ra-te* and h) is *an-i-me*.

58   TICKET TO THE MOON

A	∧	F	⌐	K	<	P	∠	U	∪	Z	⊒	-ING	:⌐
B	⌐	G	⌐	L	⌐	Q	⌐	V	∨			-MENT	:—
C	⊂	H	o	M	⊓	R	\	W	∩	CH	⊂	-ITY	:⌐
D	⊃	I	I	N	⊠	S	/	X	>	TH	∴	-NESS	:/
E	⌐	J	∪	O	O	T	—	Y	⌐	WH	⌒		

Several common words are represented by their initial letter or sound. So:

from:	⌐
have:	⊙
go:	⌐
know:	<
like:	∟
you:	⌐
the:	∴

Other abbreviations appear in the text:

letter:	∟ \
and:	⟋

Numbers are indicated by the character ⊤ , followed by the digits, with the letters from A to J representing 1 to 0.

If you know all of the above information you can translate the text into:

> FROM THE QUASAR CAME THE WIZARD CHILDREN..
> ..ERUTUF THE STCIDERP CHIWH RL A HAVE YETH
> 144 FIRE STORMS WILL SHOCK CITIES AND
> NETH ..EPACSE LLIW I AND YOU TUB .SDLEIF
> THE ZAROVO CHILD WILL BE KEPT WITH MAXI
> ..REVEROF

You might now be tempted to throw this book against a wall. It has produced some sense but also some nonsense.

The strange bracket symbols on either side of the text are the clues for how to finish the solution. One of Moon's innovations was to write lines in alternate directions: the odd lines from left to right, and the even lines from right to left. The bracket symbols help guide the finger from the end of an odd line to the beginning of an even one. This system is known as *boustrophedon*, from the Greek for "ox" and "turning," since it is how an ox ploughs a field, turning at the end of each furrow. Using a boustrophedon system, lines can be written closer together, saving space.

I included the boustrophedon style in this problem because it's so unusual, and because that's how Moon set his texts. However, computer production methods struggle with the alternate directions, so Moon type is now set with the text running from left to right on all lines. The full text is therefore:

> From the Quasar came the wizard children. They have a letter which predicts the future. 144 fire storms will shock cities and fields, but you and I will escape. Then the Zarovo child will be kept with Maxi forever.

59 THE COLOR PURPLE

The basic symbols for ColorADD are the primary colors and black and white.

| blue | yellow | red | white | black |

New colors are created by combining the symbols. Thus red + yellow = orange, and yellow + blue = green.

a) Red and yellow are shown above. Since brown = red + yellow + blue, we get:

brown

b) When the primary color/s is/are in a black square, the shade shown is the dark version of that color. And when the primary color/s is/are in a white square with a black outline, the shade shown is the light version.

Since pink = light red, and purple = red + blue, we get:

pink   dark purple

The left bracket sign means "metallic," which is why gold = "metallic dark yellow" and silver = "metallic gray."

The order is thus silver, pink, dark purple, and white.

c) Gray is a mixture of white and black, so can be written in these two ways:

light gray   dark gray

60 BUDDING BOTANISTS

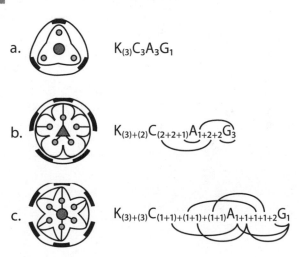

a. $K_{(3)}C_3A_3G_1$

b. $K_{(3)+(2)}C_{(2+2+1)}A_{1+2+2}G_3$

c. $K_{(3)+(3)}C_{(1+1)+(1+1)+(1+1)}A_{1+1+1+2}G_1$

The rules are:

1) For parts K, C, and A, the total number in the subscript is the number of elements. For G, the subscript denotes the number of sides the central shape has.

2) For each part, the brackets denote connections. $K_{(3)}$, for example, means that all three elements are connected, but $K_{(3)+(3)}$ means that there are two groups of three connected elements. If you need to, you can separate out what's in a bracket. So $C_{(2+2+1)}$ is the same as $C_{(5)}$. The reason you can do this is to allow you to connect the different levels, as outlined in the next rule.

3) If an element or elements on one level connect with an element or elements on another level, separate out the elements in the subscripts, and join them using arcs.

For example, in a flower with three stamens and three petals, $C_3 A_3$ shows that each stamen links to a petal. But $C_{2+1} A_{2+1}$ shows that only two stamens join to petals, and one doesn't.

LINGO BINGO

Vocab Test M–Z

1. d	5. b	9. a	13. c
2. b	6. c	10. a	14. c
3. d	7. b	11. b	
4. a	8. d	12. a	

7. Treacherous Subjects

61 WEASEL WORDS

The correct parsing of the sentence is:

The weasel [that [[a boy that startles the cat] thinks [loves smiles]]] eats.

Thus the meaning of the sentence is: The weasel eats. A boy thinks "this weasel loves smiles." That boy startles the cat.

1. "Weasel" is the subject.
2. Four: "startles," "thinks," "loves," and "eats." "Smiles" is a noun.
3. A boy startles the cat.
4. A boy thinks that a weasel loves smiles.
5. A weasel loves smiles (at least in the mind of the boy).
6. Nobody in this sentence (explicitly) smiles; "smiles" is used as a noun.
7. A weasel eats something unspecified.

62 GOOD NEIGHBORS AND GOOD FRIENDS

дӯсти	friend
хуби	good
ҳамсояи	neighbor
шумо	your

Each Tajik sentence contains the same four words, while each English sentence contains the same six words. It's likely that the two extraneous English words that do not appear in the Tajik are "a" and "of." Let's rewrite the English sentences excluding these two words. I've put the "of" part of each sentence in parentheses so the meaning is clear:

дӯсти хуби ҳамсояи шумо	good friend (your neighbor)
ҳамсояи дӯсти хуби шумо	neighbor (your good friend)
ҳамсояи хуби дӯсти шумо	good neighbor (your friend)

All three Tajik phrases end in шумо. In all three English phrases "your" appears in the second part of the sentence. It seems fair to suppose, then, that шумо = "your." It also seems likely that "your" is referring to the word/s immediately preceeding it. Thus хамсояи = "neighbor," дӯсти = "friend" and, by a process of elimination, хуби = "good." We also know that дӯсти хуби = "good friend," which tells us that Tajik word order puts adjectives (and possessive pronouns like "your") after the nouns. When we check this rule, as well as our translations in all the phrases, we see that it fits.

63 THE MOTHERS ARE WEARING THE TROUSERS

1. d
2. b

Abkhaz word order is SOV. Each Abkhaz sentence concludes in a word that ends with either *-shwup'* or *-shop'*. The most likely possibility is that this final word expresses the verb "is/are wearing." We can deduce the positions of the subject and object by looking at the patterns in which the words (and their translations) appear. You may have been initially misled by the curiosity that *anchwa* means both "the god" and "the mother."

I mentioned the three rules we need to deduce: noun plurals, verb conjugation, and the two forms of "wear."

1) Noun plurals:

	singular	plural
god	*anchwa*	
mother	*an*	*an**chwa***
squirrel	*aesh*	*aesh**kwa***
old man	*atahwmada*	*atahwmada**chwa***
son		*aba**chwa***
girl	*adzʁab*	
billy goat	*ab*	*ab**kwa***

Sometimes the plural suffix is *-chwa* (mother, old man, son), and sometimes *-kwa* (squirrel, billy goat). A reasonable hypothesis is that humans use the *-chwa* form and nonhumans the *-kwa*.

2) Verb conjugation:

I stated that the beginning of the verb changes depending on the subject. Below is a list of the subjects and their corresponding verb forms.

subject	verb
god/squirrel	*ashwup'*
old man	*ishwup'*
mother	*lshwup'*
girl	*ləshop'*
billy goat	*ashop'*
mothers	*rəshop'*
old men/sons/billy goats/squirrels	*rshwup'*

When the subject is (third-person) singular, the verb seems to take the following prefixes:

a-	nonhumans
i-	male humans
l(ə)-	female humans

When the subject is (third-person) plural, the verb always seems to take the prefix
r(ə)-

3) The word "wear":

There are two forms of the word "wear": one that ends in *-shwup'* and one in *-shop'*.

-shwup'	aprons, hats, cherkeskas, coats, shawls
-shop'	felt boots, trousers

This is the trickiest part of the problem. The different forms describe where on the body the item of clothing is worn: -shwup' is used for items worn on the top half of the body, while shop' is used for items worn on the bottom half of the body. So:

1. The son is wearing the trousers

Aba ajkwa ishop'

Since *abachwa* is "sons," we remove the *-chwa* to make the word singular. The prefix is *i-* and the verb form is *-shop'*.

2. The girls are wearing the shoes

Adzʁabchwa ajmaakwa rəshop'

The subject has the *-chwa* plural, the verb prefix is *rə-* and the verb form is *-shop'*.

64 GAH! THE GOPHER KILLED THE ANT

From looking at the first five sentences, we can deduce that the verb "kicked" must be *biztał/yiztał*, and from comparing the other sentences, we can deduce the meanings of the other words:

nouns

ashkii	boy	*naʔastsʔoọsí*	mouse
awééchíʔí	baby	*naʔazízí*	gopher
diné	man	*shash*	bear
łééchąąʔí	dog	*tsísʔná*	bee
mósí	cat	*wóláchíí*	ant
naʔashjéʔii	spider		

verbs

yiisxí/biisxí	killed
yinoołchééł/binoołchééł	is chasing
yishish/bishish	stung
yishxash/bishxash	bit

Let's get to the bottom of why the verb sometimes begins with a "y" and sometimes with a "b." Here are two of the sentences:

[boy] [man] *bitzal* = [man] [boy] *yitzal* = The man kicked the boy.

[boy] [man] *yitzal* = [man] [boy] *bitzal* = The boy kicked the man.

We can conclude that when the verb begins with a "y," the word order is subject/object/verb, or SOV, and when the verb begins with a "b," the word order is object/subject/verb, or OSV.

Now let's try to figure out what it is that makes a sentence ungrammatical, by looking at which animals appear in which sentences.

Sentences that are always grammatical irrespective of the order in which the animals appear:

man/boy

bear/baby

bee/spider

mouse/gopher

Sentences that are grammatical depending on the order in which the animals appear:

grammatical		**ungrammatical**	
1	2	1	2
boy	dog	dog	boy
baby	cat	cat	baby
man	baby	baby	man
cat	gopher	gopher	cat
bear	cat	cat	bear
mouse	bee	bee	mouse
gopher	ant	ant	gopher

You might have to stare at this list for a long time before you get the answer. In Navajo, size matters. When the noun in position 1 is the same size or larger than the noun in position 2, the sentence is grammatical. But when the noun in position 2 is larger, the sentence is ungrammatical. Bigger things must come before smaller things.

To be more specific, Navajo has a hierarchy of nouns, in which adults must precede babies, large animals precede smaller ones, and smaller animals precede insects. The complete list is as follows:

1. Human adults and children. Lightning.
2. Baby humans and large animals (bears, bulls).
3. Mid-sized animals (cats and dogs).
4. Small animals (mice and gophers).
5. Insects.
6. Plants/inanimate objects.
7. Natural forces other than lightning.
8. Abstractions.

We can now translate the sentences and indicate if they are grammatical or not:

1. The dog is chasing the baby.
2. The boy stung the bee (ungrammatical as well as implausible).
3. The spider bit the mouse.
4. The ant killed the man (ungrammatical as well as unlikely).

Since *gah* and "cat" can be used in any order, they must be on the same level in the hierarchy, so any cat-sized animal is an acceptable answer. The correct translation, as it happens, is "rabbit."

65 THE SNOB CONTAINER

As I mentioned in the build-up to the question, Kwak'wala has a system of roots and suffixes. Here are the roots used in the problem:

K'AT-	write
X̲IGW-	sweep
T'S̲AB-	dip in candlefish oil
TL̲AMK̲-	be proud/a snob
'M̲AKW-	iron
'LAG̲	make berry cakes
LIB-	play cards
KIT'L	fish (with a net)
Y̲AG̲-	knit

(I've used the roots that appear in the given examples. In some of these roots, the final consonants have mutated from the most basic form of the root. I won't go into the rules for these mutations because we don't need to know them to solve the problem. The word search format, which presents the correct spellings of all the words used, means that we don't need to deduce any mutations.)

Here are the suffixes used:

-A	turns a root into the infinitive
-AYU	the instrument used
-AKW	the result of the action
-INUXW	the expert, or person who knows how to do the action
-AT'SI	container, receptacle
-ALAS	location (you may have deduced this as "food for" which, while not strictly correct, still allows you to deduce the right words.)

Once you've worked out the roots and the suffixes, you should find several words in the word search that contain them. (A few roots have a mutated final consonant—such as a "d" changing to a "t"—but these small glitches shouldn't have put you off.) From the meanings of the roots and suffixes, you can deduce the meanings of the words:

'LAKA	make berry cakes + "A" (infinitive)	to make berry cakes
YAKA	knit + "A"	to knit
KITLA	fish with a net + "A"	to catch fish with a net
LIPA	play cards + "A"	to play cards
T'SAPA	dip in candlefish oil + "A"	to dip food in candlefish oil
K'ADAGWAT'SI	something written + container	envelope
'MAKWALAS	iron + location	wrinkled clothes
YAKALAS	knit + location	wool
YAK'INUXW	knit + expert	expert knitter
LIP'INUXW	cards + expert	expert card player

'MAGWAYU	iron + instrument	(an) iron
YAGAYU	knit + instrument	knitting needles
YAGAKW	knit + the result	something knitted
KIDLAT'SI	fish with a net + receptacle	fishing boat
TLAMGAT'SI	snob + container	tourist boat/cruise ship/ferry

Yes, a ferry is a "snob container"!

66   THE TRUTH ABOUT MIXTEC

| 1. c | 3. g | 5. e | 7. a |
| 2. d | 4. f | 6. b | |

We know that:

Nduča kaa ñíʔní   <u>The</u> water is hot

Since it's likely that the two sentences about water share the same (or almost the same) words:

Ñíʔní nduča   The water is hot

There's only one remaining sentence without a person's name, and there is only one sentence about Pedro, which leads us to deduce that:

Ndežu kaa žaʔu   <u>The</u> food is expensive
Pedro kúu xalúlírí   Pedro is my child

The statements about Maria and Juan, with their possible translations are:

Maria kúu ɨ xasɨʔɨ   Maria is a woman (or) Maria is feminine
Sɨʔɨ Maria   Maria is a woman (or) Maria is feminine

Juan kaa lúlí   Juan is my husband (or) Juan is small/short
Juan kúu xažiirí   Juan is my husband (or) Juan is small/short

It's likely that the words *kaa/kúu* are "is," so *xalúlírí* is "my child." In each of the pairs of sentences above, one is of the form "A is [noun]," and the other "A is [adjective]." We can also see that one sentence in each pair has the *xa-* form. Since *xalúlírí* is a noun, we can hypothesize that *xa-* is the noun form, which makes "woman" *xasiʔi* and "my husband" *xažiirí*. Other observations confirm that this hypothesis is correct. If *siʔi* is "feminine," and *xasiʔi* is "woman" (a female person), then it makes sense that if *lúlí* is "small/short," then *xalúlí* would be "child" (a short/small person). We can now match the sentences to their translations.

Here's a list of the grammar rules that can be spotted in the sentences above:

Rule 1: The prefix *xa-* turns an adjective into a noun with similar meaning.

Rule 2: The suffix *-rí* can be added to a noun to turn it into the singular possessive ("my" or "mine").

Rule 3: There are two ways of saying "is": *kaa* when the (predicate) phrase consists of an adjective, and *kúu,* when the (predicate) phrase contains a noun.

Rule 4: The word *kaa* is optional. When it is used, it emphasizes or stresses the subject of the sentence. When *kaa* is omitted, the word order changes.

So the answers to the remaining questions are:

i) depth	*xakǔnú*
ii) The clothes (unstressed) are red	*kwaʔá saʔma*
iii) My clothes are the red ones	*saʔmarí kúu xakwaʔá*
iv) <u>It</u> is true	*kaa ndáa*
v) It is true	*ndáa*
vi) It is the truth	*kúu xandáa*

67 DARKEST PERU

a)
1. e	3. g	5. f	7. c
2. d	4. a	6. h	8. b

b) poor   *unya*
   suffer   *nyak'ari* (*nyak'a* is fine too)
   mother   *mama*
   deer   *taruka*

c)   God is *Apu Tayta*: Spirit Father, or Father of the Spirits

68   "AND" AND "AND"

The form of "and" depends on what precedes it and what comes after it. Just like a verb, it must agree with person, number, and gender.

We can split each of the "and" words into a prefix and a suffix. The prefix is determined by the person, number, and gender of the word that comes before the "and," while the suffix is determined by the person, number, and gender of the word that comes after.

	prefix	**suffix**
I/me	*m-*	*-pa*
you	*n-*	*-cha*
he	*n-*	*-an*
she	*w-*	*-a*
they	*y-*	*-ay*

(In fact, the root of the word "and" is *a*, and the suffix is just the consonant, which either goes before or after the "a.")

1. the chieftain and you   *ncha*
2. your neighbors and I   *ypa*
3. my brother and the chieftain   *nan*
4. grandmother and my wife   *wa*
5. the guest and the hosts   *nay*
6. you and the guests' wives   *nay*
7. me and the chieftains   *may*
8. the neighbors and the guests   *yay*

Yay!

69 A CANOE PROBLEM

walini'bana vi = white person's spear = rifle/gun
ni'buna vaala = land canoe = car

Originally *vaala* meant "canoe," but when the Iatmül were introduced to cars and planes, the meaning of *vaala* started to change.

laavu	banana
laavu-ga	book (the literal meaning is "banana leaves," since sheets of paper are of similar size and shape)
laavu-ga vi'	to read

70 BAD CHILDREN HAVE SMALL UMBRELLAS

1. *Watoto wadogo wana vijoko vibaya*
2. *Mwavuli mkubwa unatosha*
3. *Visiwa vikubwa vina mito*

Swahili has more than ten noun classes, although in this question only three are used (which I'll call A, B, and C).

To approach this problem, once you compare the similarities between the words in the given sentences, you should be able to draw the following table:

	singular	plural
man	*mtu*	*watu*
child	*mtoto*	*watoto*
king	*mfalme*	*wafalme*
bag	*mfuko*	*mifuko*
umbrella	*mwavuli*	*miwavuli*
river	*mto*	*mito*
potato	*kiazi*	*viazi*
spoon	*kijiko*	*vijiko*
island	*kisiwa*	*visiwa*

From which you can extricate the prefixes for each of the three noun classes:

class A	*m-*	*wa-*
class B	*m-*	*mi-*
class C	*ki-*	*vi-*

As I mentioned in the main text, noun classes are analogous to the concept of gender in languages like French and German. Yet whereas gender divides nouns into masculine, feminine, and possibly neuter, noun classes divide nouns into types of object. For example, Class A, above, is typically used for humans, Class B for plants, and Class C for things, although there is an enormous number of exceptions.

The noun class is the most important grammatical aspect of a sentence, since it determines prefixes for both the accompanying verb and adjective. Here are the verbs used in the problem:

	class A	**class B**	**class C**
(he) has	*ana*	*una*	*kina*
(they) have	*wana*	[-]	[-]
is enough	[-]	[-]	*kinatosha*
are enough	[-]	*inatosha*	*vinatosha*

And here are the adjectives. An adjective is placed after a noun, and shares the noun's prefix.

bad	*-baya*
large	*-kubwa*
small	*-dogo*

Using all this information we can deduce the sentences 1, 2, and 3.

LINGO BINGO

Special Letters

1. a	4. a	7. d	10. c
2. b	5. d	8. b	11. a
3. b	6. b	9. b	12. c

8. Games of Tongues

71 IXNAY ON THE UPIDSTAY

Harry Potter and the Chamber of Secrets
Matilda
The Gruffalo
The Lion, the Witch, and the Wardrobe
Goodnight Mister Tom

The title of this book in Pig Latin is:
Ethay Anguagelay Overslay Uzzlepay Ookbay

The rule is take the intial consonant (or consonant cluster, such as "ch" or "pl"), move it to the end of the word, and add "-ay." If the word begins with a vowel, add "-way."

72 NAME THAT TUNE

a) Their lawyer sells the big wardrobe
b) (i) *La resolsoldo domifasol milasi rela faresimi*
   (ii) *La larefado dosoldosol fasimire la refasire soldorela*

By looking at patterns we can isolate some words, such as *faresimi* (cat) and *solmisire* (teacher).

We can also deduce that the word order is subject–verb–object, and that, just like French, the adjective follows the noun.

We know that common words have few syllables, and we see that:

redo	my
remi	your (singular)
refa	his
resol	our

In other words, the possessive personal pronouns are two-syllable words beginning with *re-*, and they follow the order of the musical scale as they progress through the singular and plural forms. (The double *rere* is missed out.) We can deduce that:

> *rela*   your (plural)
> *resi*   their

The puzzle also features the following words:

fasimire	fast	*remisifa*	slowly
laredola	buys	*ladorela*	sells
soldorela	black	*laredosol*	white

These three pairs of opposites reveal one of Sudre's curious linguistic innovations: To get the opposite of a word, you order its syllables in the opposite direction.

So, if "hates" is *silami*, then "loves" is *milasi*, and so on.

73 A FIX OF AFFIXES

1. for another reason
2. nothing
3. in every way
4. there
5. *alia*
6. *nenie*
7. *ĉiea*
8. *tiel*

You know that every *-o* word is a noun, and every *-a* word is an adjective. Here are other patterns you could have spotted. (In parentheses, I include alternative meanings that tease out the constituent concepts.)

neni-	**no**
neni-al-a	with no cause/causeless
neni-am	never (no times)

ĉi-	**all/every**
ĉi-e	everywhere (all places)
ĉi-am-o	eternity (all time)

ti-	**that**
ti-al	therefore (that's the reason why)
ti-am-a	of that time
ti-om	that many

ali-	**another**
ali-am	at another time (at a different time)
ali-el	otherwise, in another way (in a different way)

-am-	**time**
ali-am	at another time (at a different time)
ĉi-am-o	eternity (all time)
neni-am	never (no times)
ti-am-a	of that time

-e	**place**
ĉi-e	everywhere (all places)
ki-e	where? (what place?)

-el-	**way**
ali-el	otherwise, in another way (in a different way)
ki-el-o	way, mode of action

-al-	**cause/reason**
neni-al-a	with no cause/causeless
ti-al	therefore (that's the reason why)

To answer 1) to 4) we deconstruct the Esperanto words into their separate elements:

1) *alial*	*ali-al*	[another][reason]	for another reason
2) *nenio*	*neni-o*	[no](noun)	nothing
3) *ĉiel*	*ĉi-el*	[every][way]	in every way
4) *tie*	*ti-e*	[that][place]	there

To answer 5) to 8) we reverse-engineer:

5) different	[another](adjective)	*ali-a*	*alia*
6) nowhere	[no][place]	*neni-e*	*nenie*
7) omnipresent	[every][place](adjective)	*ĉi-e-a*	*ĉiea*
8) like that (in that way)	[that][way]	*ti-el*	*tiel*

74 THE HUNGRY GOAT IS TENSE

a) The goat will be going to eat/The goat will be about to eat
b) The goat has been eaten/The goat was eaten
c) *La kapro manĝantis*
d) *La kapro manĝatas*

It's easy to see that *la kapro* is "the goat," as it appears in every sentence. The stem *manĝ* is from the verb "to eat," and this question is about unscrambling the endings. Since "the goat" is the subject of all the sentences, we are only concerned with the verb's third-person singular forms.

In the list, the shortest Esperanto verb forms are *manĝas* ("eats") and *manĝis* ("ate"). It seems safe to infer that *a* denotes the present tense and *i* the past tense. (The *s* is the marker for a conjugated verb.) The other terms have endings with two parts, a first part that is either *int, ont, ant, it,* or *ot,* and a second part that is *is, as,* or *os.* If we know that *a* denotes the present and *i* the past, it would seem to make sense that *o* denotes the future. This hypothesis is confirmed when we look at the other terms with an *o* in them: They all contain some element of the future tense. Esperanto was designed for simplicity, and it's neat that a simple vowel change is the difference between the three main tenses.

In all the sentences, the goat is doing the eating, apart from 2 and 6 (*manĝitos* and *manĝotas*), in which the goat is being eaten. The former is the "active" voice, and the latter the "passive" voice. We can deduce that when the first part of the ending is *int, ont,* or *ant,* the voice is active, and when the ending is *it, ot,* or *at,* the voice is passive.

Now to determine what the two parts of the verb endings mean. Here are the verbs from 1, 5, and 9:

manĝintas	has eaten
manĝintis	had eaten
manĝintos	will have eaten

In the Esperanto, the first part of the ending has an *i*, which marks it as the past tense. In the English there is also an element that's in the past tense—the word "eaten." In these three terms, the second part of the ending contains a vowel denoting the past/present/future tense. So how do they relate to each other?

The rule in Esperanto is that the second part denotes the "current" time, and the first part indicates the tense with respect to that current time. It becomes clearer when we look at the other words in the list and their translations:

manĝ-ant-as: *manĝ*[present][present], which means the eating is being done in the present: "is eating."

manĝ-ont-is: *manĝ*[future][past], which means the eating will happen in the future relative to the past: "was going to/was about to eat."

Now to question a). The meaning of *manĝontos* is:

manĝ[future][future], which means that the eating will happen in the future relative to the future, which is "will be about to eat" or "will be going to eat."

b) The word *manĝitas* is in the passive voice, and the structure is:

manĝ[past/passive][present], which means that the goat being eaten happened in the past relative to the present, which is "has been eaten."

c) "The goat was eating" is the active voice, and the eating is going on at the same time as the past.

manĝ[present][past]: *manĝantis*

d) "The goat is being eaten" is the passive voice, and the eating is going on at the same time as the present.

manĝa[present/passive][present]: *manĝatas*

75 ■ THE WICKED GIANT ATE THE PARENTS

 a) The dwarf and the giants are speaking
 b) The taciturn (or mute) daughter is writing
 c) He loves unrequitedly (The man loves without being loved)
 d) The letter was eaten by the hungry sister

Here's a glossary of how the system works:

Nouns:

Λ̇   man   Δ̇   woman   ι̇ boy   ȧ   girl

Λ̇Δ̇   man + woman   ι̇ȧ   boy + girl
     or husband + wife   or brother + sister

Λ̇Δ̇ι̇ȧ   man + woman + boy + girl = family

⊠ Letter   ⊐⊢ Work

Family members are singled out by division and cancellation:

$$\frac{\text{Λ̇Δ̇ι̇ȧ}}{\text{Δ̇ι̇ȧ}} \quad \frac{\text{family}}{\text{(woman + kids)}} = \text{father}$$

$$\frac{\text{ι̇ȧ}}{\text{ȧ}} \quad \frac{\text{siblings}}{\text{girl}} = \text{brother} \qquad \frac{\text{Λ̇Δ̇ι̇ȧ}}{\text{ι̇ȧ}} \quad \frac{\text{family}}{\text{kids}} = \text{parents}$$

Deceased family members are preceded by a minus sign:

$$\frac{\text{ι̇ȧ(−Λ̇Δ̇)}}{(−Λ̇Δ̇)} \quad \frac{\text{kids (−parents)}}{\text{−parents}} = \text{orphans}$$

İ   person   >İ   giant   <İ   dwarf

Pronouns are composed of İ for masculine, Δ̇ for feminine, and the subscripts 1–3 for first to third person.

Plurals are denoted by the letter *n*, and the plus sign means "and." The minus sign can mean "without" or "not."

Verbs:

< talk   ⊐⊢ work   ✐ write   ♡ like/love

⌂ eat   The past tense is marked by −*t*.

Sentence structure: The subject is the base of the exponent. The exponent is the verb. The passive voice is indicated by the square-root sign. The direct object follows the equals sign.

76 | THE FACE OF BLISS

a) We know the symbols for "mouth" and "nose," and that a circumflex indicates a verb and an inverted circumflex an adjective. The symbol for "mouth" appears in four symbols, which are probably "lips," "to blow," "saliva," and "to breathe." (Do not confuse the mouth symbol with the larger circle or the circle with the dot.) The term with the mouth and the nose is likely to be "to breathe." Gradually we can piece together the meanings of the other symbols, and we realize that nouns have no grammatical markers. The symbol > is a pointer, and can be used in any orientation. (It is larger than the circumflex.)

	part of speech	composition	meaning
ô̸	verb	mouth + nose	to breathe
⌣o	noun	water + mouth	saliva
⊙̌	adjective	circle (sun) + pointer	western
⋀̌	adjective	activity	active
>◯<	noun	torso + 2 pointers	waist
ô⫸	verb	mouth + (air + outwards)	to blow
⋀̌	adjective	ill, sick	ill, sick
o̊	noun	mouth + 2 pointers	lips
ô↯	verb	eye + (water + downwards)	to weep
⋀	noun	activity	activity
♡↑	adjective	heart + upwards	merry
◯	noun	the sun	the sun

b)

	part of speech	composition	meaning
⌣	noun	water	water, liquid
⌣̆	noun	torso + pointer	neck
∨∧	verb	activity	to act, be active
⟩ŏ	noun	eye with eyebrow + pointer	eyebrow
⊕	noun	head with neck + pointer	neck

c) All the elements you need for this part have already appeared in the previous questions.

	part of speech	composition	meaning
Z	noun	air	air
◯	noun	torso	torso
↑	verb	upwards	to rise
◔	noun	circle (sun) + pointer	east
♡↓	adjective	heart + downwards	sad

77 BIG FISH LITTLE FISH

a) Today the small fish saw the fisherman.
Vera likes books.
Who will go tomorrow?

b) Yesterday Vika saw a small fisherman.

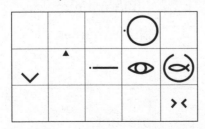

What will the reader give me?

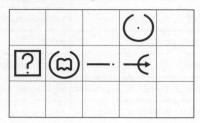

Here's a glossary of how the system works:

verbs

⟨👁⟩	to see
𝗔	to go
⊸€	to give
😊	to like

adjectives / pronouns

‹ ›	big
‹·›	many
‹|›	thick
⟷	long
›‹	small
◇	pretty
·⃞	this (thing + dot)
?⃞	who
⌈?⌉	what
⌊?	when
?	(question mark)

nouns

( )	person
(∝)	fisherman (person + fish)
(·)	me (person + dot)
∝	fish
▥	book
—	place
☐	thing
◯	day, sun
·◯	yesterday
◯·	tomorrow
└	time
·└	(in the) past
└·	(in the) future
⌷	plural marker

tenses

·—	past
—	present
—·	future

Proper names in LoCoS use a syllabary. The syllables that appear in the problem are:

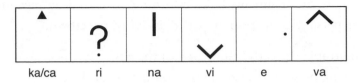

| ka/ca | ri | na | vi | e | va |

78 BLOOD OF MY BLOOD

e) *qoy, vorsa, zhav, chaf, jalan*

We're looking for the meanings of *chaf, jalan, qoy, vorsa,* and *zhav,* which are the root words that combine to make the nine longer words.

The easy root to decipher is *qoy.* Looking at the morphology of the longer words, it seems likely that *qoy qoyi* is "blood of my blood" because they both contain repeated elements.

Three other Dothraki words include *qoy,* but among the English words, the only one that might involve the idea of blood is "funeral pyre," which could be something like fire+blood.

One of the Dothraki root words is "lizard." A dragon is a lizard that breathes fire, so the Dothraki for "dragon" is presumably fire+lizard. If this is so, then looking at our Dothraki word map, the only way this permutation fits is:

qoy —— *vorsqoyi* —— ***vorsa*** —— *zhavorsa* —— ***zhav***

blood —— fire+blood —— **fire** —— lizard+fire —— **lizard**

Thus *vorsa* = "fire," and *zhav* = "lizard."

What are the other two words linked to "blood"?

A lunar eclipse is also called a blood moon. Thus *jalan* = "moon," *jalanqoyi* = "lunar eclipse," *shekhqoyi* = "solar eclipse," and *shekh* = "sun." The remaining root word, *chaf,* must be "wind," making autumn *chafka* ("season of wind") and summer *vorsaska* ("season of fire").

79 WITH AND WITHOUT

Here's how the fragments link: "Morning" is linked by meaning to "young person," and "ream of paper" is linked to "crowd."

> morning: the early/young part of the day
> young person: a person in the young part of their life

> ream of paper: a large quantity of paper
> crowd: a large quantity of people

We can now complete our semantic map:

```
        young person —— morning —— Thursday
                   |
         crowd        anthropology
           |
        ream of
         paper
```

When you draw the (morphological) map of the Láadan you will see it forms the same pattern. Each word below links only to words with a shared element.

```
        háawith —— háasháal —— hanesháal
                 |
        méwith       ewith
           |
         mémel
```

Thus:

ewith	anthropology	(science of)-person
háasháal	morning	young-day
háawith	young person	young-person
hanesháal	Thursday	(south)-day
mémel	ream of paper	[plural marker]-paper
méwith	crowd	[plural marker]-person

80 PISS AND GOLD

"Boris" is our way in. Only one of the Toki Pona terms has a capital letter, *jan Powi*, as does only one of the English words, "Boris." So it's likely they have the same meaning. I already explained that Toki Pona does not use the letters "b" or "r," which is why the word "Boris" is not a permitted word. Toki Pona replaces forbidden letters with their nearest equivalents: "b" becomes "p," and "r" becomes "w." (Other rules about translating into Toki Pona, which we don't need to go into here, mean that the "s" disappears.)

Jan Powi shares the word *jan* with two other terms. What might *jan* mean? Something, perhaps, like "name" or "person"? Looking at the English meanings, there are two other words that correspond to people: "prophet" and "robot." These seem like a good fit, so let's run with them. So *jan ilo* and *jan toki* are "prophet" and "robot," but as yet we don't know which is which.

A robot is a "machine person," and a prophet is a person who is religious/wise/senior, etc. Do any of the remaining words include the ideas of machine, or wisdom? Well, yes. A lantern is a "light machine" and a book contains wisdom, especially one that a prophet might read. So the terms *lipu toki* and *ilo suno*, which (in the map) lead from "prophet" and "robot," are likely to be "book" and "lantern," but as yet we don't know which is which.

That's the top five words in the map taken care of. Let's look at what's left: "boat," "gold," "ice," "movement," "piss," "restaurant," and "rock." Here are some ways in which the meanings are linked:

people go into them: boat, restaurant
water: boat, ice, piss
yellow: gold, piss
hard: gold, rock, ice
motion: boat, movement

It seems that "boat" has the most semantic connections to the other words. In the map, *tomo tawa telo* has the most links, so let's assume, for the moment, that "boat" = *tomo tawa telo*. What's the definition of a boat? It's a vessel for traveling over water. So it seems likely that one of the *tomo*/*tawa*/*telo* words is "water." Since "ice" and "piss" also involve water, it's likely that "water" = *telo*, since *telo* is the only one of the three terms that appears in two other words.

Thus *telo kiwen* and *telo jelo* are "ice" and "piss," but we aren't sure which is which. However, *kiwen* appears three times, once on its own, and it appears in the remaining non-water word, "gold." So *kiwen* must be the link between "gold," "rock," and "ice"; it must also be one of those words. It has to be "rock," which makes *telo kiwen* = "ice" (rock water), *telo jelo* = "piss" and *kiwen suno jelo* = "gold." We can deduce that *jelo* = "yellow." *Suno* must be something like "sparkling" or "bright," which starts to solve the problem of what the phrases for "book" and "lantern" might be. The word for "lantern" has the word *suno* in it, and the word for "book" doesn't.

We're almost done. We know that *tawa* and *tomo moku* are the remaining words, "movement" and "restaurant," but we don't know which is which. It's hard to break down movement into simpler ideas, whereas restaurant is a "house for eating." So *tawa* = "movement," and *tomo moku* = "restaurant."

You may have been helped along the way by noticing that some Toki Pona words sound similar to words in other languages, notably *jelo* ("yellow") and *toki* ("talk"). In the vocab list below I've listed the word origins for all the words.

The concepts in full are (with their etymology):

ilo	thing/tool	*ilo* (Esperanto)
jan	person	人, *jen* (Cantonese)
jelo	yellow	*yellow* (English)
kiwen	hard object/rock	*kiven* (Finnish)
lipu	flat object	*lippu* (Finnish)
moku	to eat	*mogu mogu* (Japanese)
suno	light, bright, sun	*suno* (Esperanto)
tawa	movement	*toward* (English)
telo	water	*de l'eau* (French)

| *toki* | word, to speak, to think, language | *tok* (Tok Pisin) *talk* (English) |
| *tomo* | shelter, house, vehicle | *domo* (Esperanto) |

So:

ilo suno	light tool	lantern
jan ilo	person tool	robot
jan Powi	Boris person	Boris
jan toki	word person	prophet
kiwen	rock	rock
kiwen suno jelo	bright yellow rock	gold
lipu toki	flat word	book
tawa	movement	movement
tomo moku	house eating	restaurant
tomo tawa telo	house movement water	boat
telo jelo	yellow water	piss
telo kiwen	rock water	ice

You might have deduced that *toki* = "word/language" from the name of Toki Pona itself, which means "good language."

LINGO BINGO

Pseudo Loanwords

1. c	4. b	7. d	10. b*
2. a	5. d	8. a	
3. d†	6. b	9. b	

---

* The first Russian railroad station was in the Vauxhall Pleasure Gardens in St. Petersburg, named after the (then) famous Vauxhall Gardens in London.
† In the 1950s a Japanese chef introduced the smorgasbord to Japan, but since the word "smorgasbord" is hard to pronounce in Japanese he looked for another Scandinavian word. He chose "Viking" because the film *The Vikings* was popular at the time.

9. Script Tease

81 CORNSILK AND THE DEVIL

1. j	4. i	7. l	10. a
2. e	5. c	8. g	11. d
3. h	6. k	9. b	12. f

Only two Cherokee words have two symbols (and thus two syllables): Rⱶↄ and AW. They must be *etsi* and *gola*, the only transliterations with two syllables, but we don't yet know which is which. However, three Cherokee words end in W, but only two in Ⱶↄ. We also see that three pronunciations end in *la*, and two in *tsi*. It's looking likely that Cherokee reads left to right, and that W = *la* and Ⱶↄ = *tsi*, which would mean that AW = *gola* and Rⱶↄ = *etsi*.

The other Ⱶↄ (*tsi*) word is Oⱽⱱ9Ⱶↄ = *uwetsi*, whose first syllable Oⱽ must be *u*, and thus Oⱽⱶↄ⊖, the only other word beginning with Oⱽ is *uyona*. The Ⱶↄ (*tsi*) syllable appears in two other words, DⱶↄW and Dⱶↄ16ↄ0У, which must be *atsila* and *atsilvsgi*, based on the number of syllables. We can now state that the other word ending in W (*la*) is *adela*.

Three words end in Ⱶ, which must be *di*, since three pronunciations end in *di*. 4M OⱽꞀⱶↄ9ↄ is two words, so it is *selu unenudi*. We know that A = *go* (from AW = *gola*), so AⱵGⱽꞀ = *gohiyudodi* and GↄꞀⱤbↄ = *tsatiyosidi*.

To finish, we know that ⊖ = *na* from *uyona*, so Cↄↄ0У⊖ = *tsvsgina*, and the word that remains, SↄZⱵↄi6ↄ0E, must be *ganolvvsgv*.

82 A COOL CALENDAR

LⱵ	*mai*
Dↄↄↄ∧⌐	*utupiri*
∩ʸ∧⌐	*tisipiri*

The system is syllabic, meaning that there is a symbol for each syllable. By looking at *sitipiri* and *nuvipiri*, which each have four syllables and four symbols, we can deduce the symbols for *si*, *ti*, *pi*, *ri*, *nu*, and *vi*, which gives us enough information to write *tisipiri*.

By looking at the other words, you may deduce the following rules:

1) A dot above a symbol adds an extra vowel in the transliteration.

2) When a syllable ends in a consonant, that consonant is represented by a symbol in superscript.

Here's a grid of the symbols that are used in the problem, and their corresponding pronunciations:

	a	i	u	(superscript)
	ᐊ	ᐃ		
p-		ᐱ	ᐳ	
v-		ᕕ	ᕗ	ᕝ
n-		ᓂ	ᓄ	ᓐ
j-	ᔭ		ᔪ	
l-	ᓚ			
m-	ᒪ			
t-		ᑎ		ᑦ
g-		ᒋ		ᒡ
r-		ᕆ		
s-		ᓯ		

The idea behind the system is that the orientation of a symbol denotes the vowel sound in the syllable. The triangle pointing left is "a" and the triangle pointing up is "i." The question requires you to deduce that a triangle pointing right must be "u." You may have come to this conclusion by noticing that the syllables *pi* and *vi* are pointing up, whereas the syllables *pu* and *vu* are pointing right. And if *ti* is pointing up, *tu* must be the same shape pointing right.

Eagle-eyed readers may have also deduced the (correct) hypothesis that the superscript symbol is a small version of the relevant consonant with the vowel "a." If you consider the symbol for the consonant with the vowel "a" as the basic form, then to transform the vowel sound to a "u" you flip the basic form across the vertical axis.

Inuktitut symbols only require three orientations because the language only has three vowel sounds—"a," "i," and "u"—which can be either short or long. The long versions have dots above the relevant symbols, which correspond to double vowels in the transliterations.

83 DESTINATION TIMBUKTU

1. k	4. d	7. j	10. g
2. h	5. a	8. b	11. c
3. e	6. i	9. l	12. f

N'ko is ⵗⵂⴴ, and Kanté is ⴷⴱⵗⵂ.

(First, note that the horizontal tail in most of the letters is the cursive stroke that joins them. This line is not present on the right side of any of the words, which may have been a clue about the right-to-left direction of the writing.)

The symbol ⵂ appears in both N'ko and Kanté, so must be a "k." (When letters are printed individually they do not include the horizontal "joining" line.) We can deduce that the script is read from right to left. This isn't so surprising: Arabic reads from right to left, and we know that one of Kanté's motivations in designing the alphabet was to transcribe the Koran. In fact, Kanté decided on the direction of the script after asking all the people he knew to draw a line in the sand. Ninety percent drew it from right to left.

Two transliterated names begin with a "k," and two in N'ko (5 and 8) begin with a ⵂ. One of them, *konakri*, has two "k"s. Since 5 has two ⵂs, *konakri* must be 5 and *kindiya* 8.

We can also start to make some other likely correspondences:

k	=	ⵂ	a	=	ǀ
ɔ	=	ⵄ	r	=	✝
n	=	ⵗ	i	=	Y

Now that we have the letter for "a," we can deduce that word 6 must be *abijan*, and that the dot at the bottom of the final "a" indicates an "n."

From the word N'ko, we know that *n'* is ꧋. Two words, 7 and 11, begin with this, so we can assume that one is *n'srégbdɛ,* and the other *npraeso*. With the letters we think we have identified, 7 has the combination *n_ra__o.* (the final "o" we get from N'ko), and 11 has *n_r_____*. It's looking like 7 is *npraeso,* and 11 is *n'srégbdɛ,* and this is confirmed when we consider that 11 also has the symbol ꭉ. This symbol also appears in 8 (*kindiya*), from which we deduce that it is probably a "d," since in *kindiya* the ꭉ is preceded by a Ɣ, which is probably "in" since we hypothesized that Ɣ is "i" and a dot at the bottom is an "n."

These two provide us with new letters with which to identify other words. For example, if "s" is ◻, then word 4 is *sromaya,* and 1 must be *gesoba,* meaning that the letter ∇ is "g," and word 2 must be *bisawo*. From the letters we already know, word 3 must be *faranna* because it's *?-a-r-an-n-a,* and word 9 must be *gkedu* because of the "d." The initial letter of 12 is the same as the "j" from *abijan,* and so 12 must be *jikuɛ*. And thus, by a process of elimination, the one remaining word, 10, is *tonbtu,* aka Timbuktu. (Which we could have deduced anyway, since it repeats its first letter, and also contains a "b.")

84 TWISTING TWI TWISTERS

1. g	3. d	5. h	7. c
2. e	4. f	6. a	8. b

The animals in f) are two crocodiles. The tool in a) is a wooden comb. Akan people associate women with the moon, hence the moon symbol in a) and b). *Osram ne nsoromma* means "moon and star," but give yourself a point if you thought *nsoromma* meant "sun." The idea is that the moon only receives its radiance from the sun, thus the foundation of success is mutual cooperation.

85 GEORGIA ON MY MIND

1. Argentina
2. Colombia

The words for "Uruguay" and "Peru" have the same number of letters in Georgian and in English. Usually the names of countries sound broadly similar in all languages, so let's assume a one-to-one correspondence between letters for each language. The fact that the word "Uruguay" has the same letter in first and third position means that it is likely that Georgian script reads from left to right. The word "Brazil" uses the "r" and "a" letters from the other words, so it looks like the letters at the beginning of the word at least match those in the English translation, even if the Georgian has eight letters rather than six (this is because the transliteration of the Georgian word is "Brasilia"). The letter ი corresponds to "i" or "y."

Filling in the known letters in არგენტინა gives:

A R G E _ _ I _ A

And in კოლუმბია gives:

_ _ L U _ B I A

The question only concerns South American nations, so it makes sense that the answers will too. Once you realize this, your success in deducing the solution depends on your geographical general knowledge.

86 A YEAR IN YEREVAN

January	հունվար	July	հուլիս
February	փետրվար	August	օգոստոս
March	մարտ	September	սեպտեմբեր
April	ապրիլ	October	հոկտեմբեր
May	մայիս	November	նոյեմբեր
June	հունիս	December	դեկտեմբեր

Right away we see that there are two types of repeated endings:

-Եմբեր, as in Հոկտեմբեր, սեպտեմբեր, դեկտեմբեր, and նոյեմբեր

and:

-վար, as in Հունվար and փետրվար

The months in English have similar repeated endings. So we can assume that the top four are the "-ber" months, and the bottom two are the "-uary" months. Also, of the months that are not "-ber" months, three begin with Հ and two with Մ, which again is similar to the months in English: Three begin with J, two begin with M, and two begin with A. If we assume that Հ is "j" (it is actually an "h"), then January must be Հունվար, and June and July must be Հունիս and Հուլիս, although we don't yet know which is which.

Looking at the endings, February must be փետրվար. The word for January has the character Ն, which is also the first letter of նոյեմբեր, so we can be fairly confident that Ն is "n" and this month is November. The "n" also appears in one of the J months, which we can deduce must be June, so the other is July.

Looking at the lengths of the words, it seems more likely that March and May, rather than April and August, are the Մ-months. We also notice the letter ր in our "-uary" and "-mber" suffixes. It's likely that this is a "r" (which it is), so we can deduce which M-word is March and which is April (since they both have a ր). We get August and May for free. The պ in April, probably a "p," gives us September. It's a toss-up between the remaining two, October and December, but comparing the first letters with November makes us think that ո is probably "o" (it is), and we are done.

87 SORCERY IN SOUTHEAST ASIA

1. e	3. h	5. d	7. b	9. g
2. i	4. a	6. f	8. j	10. c

We see that 6 begins with a repeated character, ꦧꦩ, so it must be *babad*, with ꦧꦩ = *ba*. (The last part, ꦧꦢ, is composed of a ꦢ = *da*, and a ꧀, which supresses the final vowel sound, although we don't need to be too concerned with that here.) The word *jawa* has two syllables, both with the vowel "a," so we know that it must consist of two Javanese consonants with no diacritics. The simplest Javanese word is 3, which is composed of two letters, ꦗ and ꦮ, with no diacritics. So 3 must be *jawa*, and thus ꦗ = *ja* and ꦮ = *wa*.

What makes Javanese hard to decipher for neophytes—but also what makes it a particularly beautiful script—is that the other vowels attach to the consonants in a way that distracts from their shape. Thus, we see that ꦧꦩ and ꦗ appear in other words in disguise. For example, the first letter in 1 is ꦗꦸ, which is ꦗ with a descender for the vowel "u."

Thus 1 is *juruh*. ꦗ also appears in 4, which must be *ngajar*, with the accent adding an "r" sound to the vowel. Looking through the words, we find a ꦧꦩ = *ba* in 7, so it must be *borang*. It seems odd that the "b" sound is not the first character. This peculiarity arises because the diacritic for "o" has two parts, a ꦧ before the letter and a ꦴ afterward.

Moving on, since we know the shape of the "u" descender, we can deduce that 8 must be *yumani*, and that 10 is *teluh*. The "m" in *yumani* is ꦩ, which means that 9 must be *malikat*. The symbol that looks like a voice bubble emerging from 8 and 9 is the vowel "i," which means that 5 must be *kesiring*. The remaining word, 2, must be *pentil*.

You may have noticed that *pentil* is the only word that has two consonants together, the "n" and the "t." (The "ng" in *borang* and *kesiring* is a single consonant sound.) It is also the only word written in Javanese in the puzzle that has one consonant on top of another—which is how double consonants are written, and is another clue to solving this (hard!) problem.

88 BLAME IT ON THE BUGIS

B, E, D, I, H, A, C, J, G, F

I warned you this was a toughie! Lontara has no spaces between words, which makes it difficult to read.

Let's start by reading the transliterations. Lines E and F are identical: *mattampa puang lé ri batara*. Looking at the English translation, the only phrase used twice is "to call the gods Lord," which appears near the beginning of the text (followed by a comma) and at the very end (followed by a period). We can deduce that *mattampa puang lé ri batara* means "to call the gods Lord." Our first task is to locate this repeated phrase in the Lontara text. We know it appears at the end of the text, so by working backward from the final character, we can isolate a string of characters that is repeated near the beginning.

This repeated text is the whole of the bottom line, and a sequence beginning with the third-last character on the first line:

∨∧⋙∧̣ ᴍ⟨ᴇ̃ ᷄⋩∧❀

We know Lontara is an alphasyllabary, and that this sequence (of 11 characters) is the phrase (of 10 syllables):

ma-tta-mpa pu-ang lé ri ba-ta-ra

(The reason there are 11 characters but only 10 syllables is because one of the characters, the ❮, is a diacritic. We don't need to know that to answer the question.)

The second and penultimate symbols in the phrase are the same, as are the second and penultimate syllables. So:

∧   *ta*

We can deduce that the first symbol/syllable is:

∨   *ma*

Continuing with the third symbol/syllable, we get:

⋙   *mpa*

And from the end of the word, we get:

❀   *ra*

⋩   *ba*

The fourth-to-last symbol is the same as the last symbol, apart from the extra dot Ȯ. The fourth-to-last syllable is *ri* and the last syllable is *ra*. Based on what we know from alphasyllabaries, we can deduce that the basic form of the character is an -*a* sound, and that the Ȯ dot on top changes the vowel to an -*i*.

The fourth symbol and syllable are:

    ᴧ̣   *pu*

We can deduce that the diacritic is the ꞈ, which denotes the vowel -*u*, so:

    ᴧ   *pa*

From our rudimentary syllabary, we can start to deduce other parts of the text:

    ᐊᴧᴧ *ma* ꞈᴧ *mu* ᴧᴧȯᴧ *ma-tta-mpa*
    pu-ang lé ri ba-ta-ra ⸫ *ma pa* ᴧ̀
    ᴧᴧ̀ *ri* ᐊ *pa* ᐊ *ra ti* ᴧ̈⸫ *ta ma* ᴧ
    pu ᴧ *mu* ᴧᐊᐸᴏᴧ̣ᴧ *ri* ᴧȯᴧ *mu* ⸫
    ta ba ᵃ̀ᴀ *ba* ᵃ̀ᴀ *ri* ᴧ *ta* ᴧᵃ̀⸫
    ᴧᴧᴧᴧᐱᴧᵛᴧᐱ *ba* ᴧ̀ᴧᴧᐊᴧ
    ma ꞷ *ta* ᴧᵌ *ri* ᴧ *ta* ᴧᵃ̀⸫
    ᵌᴧᴄꞷᴧ *ta* ᴧ⸫ *pu* ᴧ⸫ᐊ *ra* ꞷᴧ
    ma ꞈᴧ *ta* ᴧᵞ *ri* ᴧᴧᴧᵼ̀⸫
    ᐊᴧ *ri* ᐊ *ma* ᐊᴧᴧᐊ *pa* ᐊ *ra ti* ᴧᐊᴧ⸫
    ma-tta-mpa pu-ang lé ri ba-ta-ra ⸫

The first line has a *ma* and a *mu* separated by two syllables. In the transliterations, the only place this happens is B, so this must be the first piece of text.

The only time we have *ma* followed by a *pa* is in D, so D must be the third snippet, and the three dots a punctuation mark. D ends with a punctuation mark, and immediately afterward are the syllables *ta* and *ma*, which must be I. The line after that begins *ta-ba*, which must be H. By continuing to fill in the gaps in this way, we can deduce the correct order. (The reason some syllables appear to have the wrong vowels is because they have diacritics we have not yet identified.)

89 THE STROKES

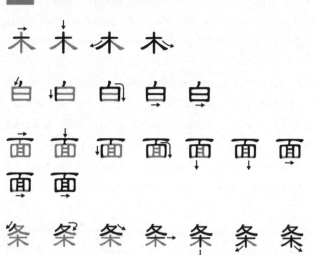

The overall principles seen in this question are:

1) Horizontal strokes before vertical ones
2) Left-falling strokes before right-falling ones
3) Top to bottom
4) Left to right
5) Outside to inside
6) Frames are closed last

90 GANGNAM STYLE

1. d
2. a
3. c
4. b
5. 삼성 = Samsung

Notice that each of the Hangul words has two blocks, and each of the names in English has two syllables. Each block represents a syllable, and the letters go from the top to bottom in each block. The word 평양 has the ㅇ symbol at the bottom of both blocks. The only word in which both syllables end with the same sound is Pyongyang, so ㅇ is an "ng," and 평양 = Pyongyang. We can see the "ng" in 강남, which must be Gangnam, from which we can deduce that Hangul reads from left to right, and that ㅁ is "m." (The shape one's lips make when pronouncing "m.")

Vowels are either a horizontal or vertical line with one or two minor lines off it. From Gangnam, we can deduce that ㅏ is the vowel "a." Which means that 부산 must be Busan (*Bu-san*), so 서울 is Seoul (*Seo-ul*). Thus, ㅜ and ㅓ are the vowels "u" and "eo," and ㅅ is the consonant "s." We now have all the elements we need to deduce 삼성, which must be *Sam-seong*. If you got this answer, award yourself full marks.

Somewhere along the line the Korean company whose name is pronunciated "Samseong" decided to ignore the subtle pronunciation difference between "eo" and "u," and that its brand name in Latin spelling would be "Samsung." A non-Korean person trying to pronounce a Korean "eo" is going to say something like "u" anyway.

Schubert's *Die Forelle* was a clue: It is the melody played by Samsung washing machines when they complete a cycle.

LINGO BINGO

Untranslatables

1. a	4. c	7. c	10. a
2. c	5. b	8. b	
3. a	6. d	9. b	

10. Oh My Days

91 ■ SNOW TIME

15:55   *Diibmu lea vihtta váile njeallje* ("five to four")
16:20   *Diibmu lea logi váile beal vihtta* ("ten to half before five")
18:35   *Diibmu lea vihtta badjel beal čieža* ("five past half before seven")
22:10   *Diibmu lea logi badjel logi* ("ten past ten")

The main difference between telling the time in Sami and in English is that the Sami use the phrase "half before" the next hour, whereas in English it's "half past" the previous hour. Here's one way you might have gotten there.

The phrase *diibmu lea* appears in every line, so it probably means something like "the hour is . . . ." We don't need to know its exact translation to solve the problem. The word *okta* appears in both 13:10 and 12:30. In what way might the same word be included in these two times? Well, 13:10 is ten past one, and 12:30 is half an hour before one. So let's assume that *okta* is "one," which would mean that *logi badjel* is "ten past," and *beal* is "half before." From *vihtta badjel čieža*, five past seven, we could deduce that *vihtta* = "five," *badjel* = "past," and *čieža* = "seven." This is beginning to make sense. *Logi váile vihtta* is ten to five, confirming that *logi* = "ten," so *váile* = "to."

Since 12:30 is written as "half before one," it's logical to assume that 9:30 would be written as "half before ten," or *beal logi*. Which is why 9:25 is written as "five before half before ten." And 3:40 is written as "ten after half before four," giving us *njeallje* = "four."

92 A WEEK IN TOKYO

Sunday	*Nichiyobi*
Monday	*Getsuyobi*
Tuesday	*Kayobi*
Wednesday	*Suiyobi*
Thursday	*Mokuyobi*
Friday	*Kinyobi*
Saturday	*Doyobi*

The question tells us that the Japanese words will be based on the sun, the moon, fire, water, wood, metal, and earth. So something like sunday, moon-day, fire-day, water-day, wood-day, metal-day, and earth-day. We know that Thursday is wood-day, and by looking at *Mokuyobi* and *Mokuhanga*, we can deduce that *Moku* is "wood," and that *yobi* is "day." Likewise, "water" is *sui*, "fire" is *ka*, "Earth" is *do*, "Moon" is *getsu*, "Sun" is *nichi*, and "gold/metal" is *kin*.

93 TELLING THE TIME IN TANZANIA

1. d	4. e	7. a
2. g	5. f	8. b
3. h	6. c	*jumatatu* = Monday

When we look at the original list, we see that every phrase begins with a *j*-word before a comma. It's likely, then, that the *j*-words are the days of the week. But we run into something odd right away. In the English expressions we see 3 × Sunday, 2 × Tuesday, and 3 × Saturday. But in the Swahili there are 4 × *jumamosi*, 2 × *jumanne*, and 2 × *jumapili*. Clearly, Swahili days do not correspond precisely to English ones.

Four of the Swahili phrases end in *usiku*, and four end in *asubuhi*. This could be their version of "AM" or "PM." The remaining part of each expression is of the form *saa* X, *saa* X *na robu*, or *saa* X *na nusu*. This is probably the number. In fact, we can assume that *robu/nusu* are 15/30 or "quarter"/"half," but we don't yet know which is which. The only "round number" times are 1:00 and 7:00, and the only Swahili expressions written *saa X* are *saa moja* and *saa saba*. So it's likely that 1 and 7 are *moja* and *saba*, but again, we don't know yet which is which.

Let's assume that *saba* = 1 and *moja* = 7. This gets us:

Sunday 1 AM = *jumamosi, saa saba usiku*

Saturday 7 PM = *jumamosi, saa moja usiku*

Sunday 7:30 AM = *jumapili, saa moja na nusu asubuhi*

Hmm. From this we deduce that *jumamosi* = "Saturday" and "Sunday," and *jumapili* = "Sunday," which seems like a contradiction, since one day corresponds to two. Although if we look at the times for Saturday and Sunday when they are *jumamosi,* we have Saturday 7 PM and Sunday 1 AM, both of which are *usiku.* And *jumapili* is Sunday at 7:30 AM, which is *asubuhi.* Which leads us to think that maybe one day does not turn into the next at midnight, but sometime between 1 AM and 7:30 AM. And that *usiku* is "nighttime" and *asubuhi* is "daytime."

Let's run with that. It would mean, therefore, that Sunday at 9:15 AM is *jumapili* and *asubuhi,* which corresponds to the expression containing *tatu* and *roba,* which we can deduce must be "9 and 15/one quarter."

The two remaining *jumamosi* are *saa mbili na robu usiku* and *saa nne na nusu asubuhi.* Our options are Saturday at 10:30 AM and 8:15 PM, so *mbili* = 8 and *nne* = 10.

We're left with the *jumanne* expressions, which must be Tuesday. So *tano* = 11 and *sita* = 12. This gives us the following times correctly matched to their Swahili translations:

Sunday, 1:00 AM	*jumamosi, saa saba usiku*
Sunday, 7:30 AM	*jumapili, saa moja na nusu asubuhi*
Sunday, 9:15 AM	*jumapili, saa tatu na robu asubuhi*
Tuesday, 12:15 PM	*jumanne, saa sita na robu asubuhi*
Tuesday, 11:30 PM	*jumanne, saa tano na nusu usiku*
Saturday, 10:30 AM	*jumamosi, saa nne na nusu asubuhi*
Saturday, 7:00 PM	*jumamosi, saa moja usiku*
Saturday, 8:15 PM	*jumamosi, saa mbili na robu usiku*

Well done if you got this far. You've got the correct answer, even if we made a mistake on the way and there are still some ambiguities to figure out. We'll clear everything up when we get to the final question.

According to our workings, we have these numbers matched:

1	*saba*
2	
3	
4	
5	
6	
7	*moja*
8	*mbili*
9	*tatu*
10	*nne*
11	*tano*
12	*sita*

In the question, we learned that there is a day called *jumatatu*. In other words, *juma*-9. If we look at the structure of the other days of the week, we have *juma* + *mosi*, *juma* + *pili*, and *juma* + *nne*. The first two would be pronounced in a similar way to *juma*-7 and *juma*-8, and the third is exactly *juma*-10.

It seems strange that the numbers 7, 8, 9, and 10 should appear in the words for days of the week, when a week has only seven days. So we must have made a mistake somewhere.

Our error is now clear. We assumed that *saba* = 1, and *moja* = 7. But what if we assumed the converse? The order of the dates stays the same, but our Swahili numbers are now *moja* = 1, *mbili* = 2, *tatu* = 3, *nne* = 4, *tano* = 5, *sito* = 6, and *saba* = 7.

Thus, the Swahili week starts on a Saturday, and the third day of the week, *jumatatu*, is Monday.

Equipped with the correct number translations, we can get a better understanding of the Swahili expressions of time:

7:30 AM is actually "1:30 AM"

9:15 AM is actually "3:15 AM"

And so on.

In other words, in Swahili the days change not in the middle of the night, as they do for English speakers, but at 6 AM, at sunrise.

94 EVIL DAYS

Look at the patterns in the given words: Some start with the same letters; some end with the same letters.

The two words that start with *Kuru-* are 6 days apart.

The five words that start with *Mono-* are 12, 6, 12, and 6 days apart. (12 is a multiple of 6.)

The two words that start with *Nkyi-* are 30 days apart. (30 is also a multiple of 6.)

The two words that end in *-ya* are 7 days apart.

The two words that end in *-dwo* are 14 days apart.

The two words that end in *-kuo* are 21 days apart.

The two words that end in *-kwasi* are 35 days apart. (14, 21, and 35 are multiples of 7.)

We know that the 42-day cycle is a combination of a 6-day and a 7-day cycle. We can deduce, therefore, that the beginning of a word marks the 6-day cycle, and the end of a word denotes the 7-day cycle.

Working out the 6-day sequence, we get:

Fo-, [not given], *Nkyi-*, *Kuru-*, *Kwa-*, *Mono-*

And the 7-day sequence is:

-dwo, -bena, -wukuo, -ya, -afi, -mene, -kwasi

So:

  a) *Fodwo*
  b) *Kuruwukuo*
  c) *Foafi* (in fact, this day is shortened to *Fofi*)
  d) *Kurukwasi*

95 A YEAR IN ADDIS ABABA

መስከረም	*Mäskäräm*	ሚያዝያ	*Miyazya*
ጥቅምት	*Ṭəqəmt*	ግንቦት	*Gənbot*
ኅዳር	*Ḫədar*	ሰኔ	*Säne*
ታኅሣሥ	*Taḫsaś*	ሐምሌ	*Ḥamle*
ጥር	*Ṭərr*	ነሐሴ	*Nähase*
የካቲት	*Yäkatit*	ጳጉሜ	*Ṗagume*
መጋቢት	*Mägabit*		

First we need to work out the direction of the script. The symbol ት appears four times as the rightmost symbol of a word. In the transliterations, letter "t" appears four times as the rightmost letter of a word. Since no other letter appears four times at either end of a word, it's reasonable to suppose that "t" = ት, and that the script reads from left to right.

Two words begin with መ, and two with ጥ. We know that these are syllables. Two words begin with *Mä: Mäskäräm* and *Mägabit*, and two with *Ṭə: Ṭəqəmt* and *Ṭərr*. Looking at the lengths of the Amharic words, a fair guess would be that ጥር is *Ṭərr*, and ጥቅምት is *Ṭəqəmt*. *Mäskäräm* begins and ends with an "m," which suggests that it is መስከረም, which contains a መ on the left and a variation of መ on the right. This leads us to conclude that *Mägabit* is መጋቢት and *Miyazya* is ሚያዝያ. The shortest remaining transliteration is *Säne*, which is likely to be the shortest remaining word in Amharic, ሰኔ. The symbol ሐ appears at the beginning of one word and in the middle of another, suggesting that this is "ha." Thus *Ḥamle* is ሐምሌ, and *Nähase* is ነሐሴ.

96 TZOLK'IN ABOUT A REVOLUTION

Let's first concern ourselves with the glyphs on the right, the ones that look like ornate signet rings. Three of these appear more than once:

 appears on August 18, September 7, and September 27: two gaps of 20 days.

appears on August 15, September 4, and September 24: two gaps of 20 days.

 appears on August 13, and September 22: a gap of 40 days.

We know that every right-hand glyph repeats after the same number of days, so it must repeat at least every 20 days. In fact, the calendar has examples of different right-hand glyphs that are 2, 4, 5, and 10 days apart, so we can eliminate the possibility that the repeat number is any of 2, 4, 5, or 10, since if the repeat number was 2, 4, 5, or 10 then any two glyphs this number apart would be identical. So every right-hand glyph repeats after 20 days.

Now what about the glyphs on the left?

The glyph ⦂ appears on August 6, August 19, and September 27: a gap of 13 days and a gap of 39 days. (Note that this glyph is not the same glyph as the ones on August 16, 29, and September 24, which include extra bars.) The number 13 has no divisors, so it must be the case that left-hand glyphs repeat after 13 days. We can confirm this by checking the gaps between other reappearing glyphs. Also, if the left-hand glyphs repeat every 13 days, and the right-hand ones every 20 days, then every combination of right- and left-hand glyph must repeat every 13 × 20 = 260 days, as stated in the question. So we know we're on the right track.

Now we need to find the correct left- and right-hand glyphs for September 28, 2007.

The one on the right will be a glyph that appears either 20 or 40 days before that day, in other words, ⊙, which appears 40 days before, on August 19.

The left-hand glyph requires more work, since none of the dates that are multiples of 13 days before September 28 are marked on the calendar.

Let's assemble all the left-hand glyphs, and put them into the order in which they appear in the 13-day cycle, letting day 1 = August 1:

Wait! We also have two more left-hand glyphs available that were set in the question. These are just like the two in positions 5 and 6 above but with a vertical bar. It would seem logical that they fit in spaces 10 and 11 in the same order. Hence:

It should be no surprise that the glyph we require for September 28 is the missing one. But what can it be? Let's look again and try to spot a pattern. If these are numerals, where does the sequence start? Suppose we move the three most complex from the start to the end:

This looks promising: The number system seems to be small circles for units, and bars for fives. Which would make the missing number 4, consisting of four white circles:

1	2	3	4	5	6	7	8	9	10	11	12	13

The "c"s in the glyphs for 1, 2, 6, 7, 11, and 12 are purely ornamental, there to fill vacant spaces.

So the answer to 1) is:

2. a) The Mayan number is 7, so the date must be one of:
        August 10, August 23
        September 5, September 18

The date can be either August 10 or 23, since the other two dates are 20 days from a different right-hand glyph. We can only find the correct answer once we have done part b), since this eliminates the possibility of August 10. So the answer is August 23.

b) The Mayan number is 8, so the date must be one of:
August 11, August 24
September 6, September 19

The date must be September 19, because all the other dates are 20 days from a different right-hand glyph.

97 THINKING INSIDE AND OUTSIDE THE BOX

3. *Wǒ yào chūqù*   I want to go out
4. *Tā jìnqùle*   He/she went in
5. *Tā yào jìnqù*   He/she wants to go in
6. *Tāmen yào jìnlai*   They want to come in
7. *Tāmen chūqùle*   They went out
8. *Tāmen chūláile*   They came out
a) *Tāmen yào chūlái*
b) *Tā jìnláile*

We can make the following deductions by looking for repeated words and elements:

Whenever the arrow goes to or from a black dot, the pinyin contains *wǒ*.

Whenever the arrow goes to or from a single white dot, the pinyin contains *tā*.

Whenever the arrow goes to or from two white dots, the pinyin contains *tāmen*.

The black dot is the speaker, so *wǒ* = "I" (which we can confirm in the English translations), and it would appear logical that *tā* = "he/she" and *tāmen* = "they." (In Chinese, the words for "he" and "she" contain different characters but sound the same.)

When the arrows are pointing out of the box, the prefix *chū-* is used, and whenever the arrows are pointing into the box, the prefix *jìn-* is used. So we can conclude pretty safely that:

chū-   out
jìn-   in

Now, look at whether the circles are at the tip or the tail of the arrow. When they are at the tail, the word *yào* is included, but when they are at the tip, the second word has the suffix *-le*. Looking at the translations of the first two sentences, and considering the intuitive meaning of being at the tip or the tail of an arrow (of time), it would seem that with *yào* the action is about to happen, and with *-le* the action has happened. So:

> *yào*   want(s) to
> *-le*   [past tense]

The remaining elements are *qū* and *lái*. These depend on the position of the black circle, the symbol that denotes the speaker. When the arrows point *to* the region where the speaker is, the *lái* appears. But when the arrows point *from* the region where the speaker is, the *qū* appears. So:

> *qū*   indicates movement away from the speaker
> *lái*   indicates movement towards the speaker

This gives us the translations listed above.

a) The picture shows two people wanting to go out of the box toward the speaker. In English that would be "They want to come out," and in Chinese it is:

> *Tāmen yào chūlái*

b) Here we have a person having arrived in the same region as the speaker, which in English is "He/she came in," and in Chinese is:

> *Tā jìnláile*

98 A MALTESE TEASER

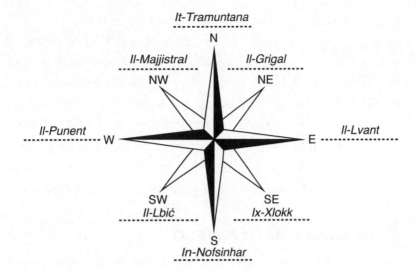

The initial letters of the mnemonics give us

    T-L-N-P
    G-X-L-M

which we can assume stand for the winds. Most probably, one group of four will be the cardinals, and the other will be the intercardinals, and there will be a way to list them clockwise (or counterclockwise).

The two winds derived from the names of countries are: *Il-Grigal* (Greece) and *Il-Lbiċ* (Libya). The wind that has a name of a region in the Middle East is *Il-Lvant* (the Levant). By counting two positions from L in each mnemonic, we can deduce that *Il-Lvant* is opposite a wind beginning with either G or P. It can't be *Il-Grigal*, since we know this wind is not derived from the name of a country, so it must be *Il-Punent,* which has a Latin root (*ponere,* "to put"). The Latin-sounding name *Il-Tramuntana* must be the cold wind that comes from high land. Another clue is that it has a root that sounds like "mountain."

We can thus position the winds in the mnemonic:

*Il-**Tramuntana** – Il-**Lvant** – In-**Nofsinhar** – Il-**Punent***
*Il-**Grigal** – Ix-**Xlokk** – Il-**Lbiċ** – Il-**Majjistral***

Now it's just a case of lining up them by looking at the map.

From Malta, the Levant looks to be east and Greece northeast. *Nofsinhar* is a time of day, so probably something to do with the sun. (It means "midday.") North of the equator, the sun is always south, so it's a good bet that *In-Nofsinhar* is due south. Also, cold winds are likely to come from the north, and directly north of Malta is (hilly) Sicily, so this would seem to confirm that *Il-Tramuntana* is north. Of the intercardinals, Greece must therefore be northeast and Libya southeast, and once these two are in position the others follow.

99 HUNGARIANS IN A FIELD

From the translation we are given, we know that *egy kő van* = "there is one stone." Each of the Hungarians (who we will call A, B, C, and D) makes a statement that includes the phrases:

> *egy kő van*
> *két kő van*
> *nincs kő*

Looking at the numbers of stones in any line in the grid, it's likely that *kő* = stone, and that *két* = 2 and *nincs* = 0.

If every position occupied by a Hungarian has no stones in one cardinal direction, one stone in another direction, and two stones in a third, we can already mark the only possible positions of the Hungarians in the grid, which are numbered below:

We know that the cardinal directions are:

keletere   east
délre
északra
nyugatra

And that the other words are:

mögöttem behind me
balra
elöttem
jobbra

The fact that we know A's position, and that we also know in which direction from A there are two stones (south) and no stones (in front of A, and to the right of A), means that *délre* must be south and *jobbra* either "in front of me" or "to the right of me." However, the word *elöttem* is very similar to *mögöttem*, so it's likely that this word means "in front of me." So *jobbra* is "right" and the remaining word, *balra*, is "left."

The words *északra* and *nyugatra* must be "west" and "north," but as yet we don't know which is which.

Let's look at what C is saying:

"*Északra* (in front of me) there are no stones"
"*Nyugatra* there is one stone"
"To the right of me there are two stones"

This establishes that there are no stones in front of C, but two stones to the right. This eliminates positions 4 and 7, since in both these positions there is no direction that C can face such that no stones are in front them and two are to the right. If C were in position 1, the constraints would force C to face east, making *északra* = "east," which is forbidden since *keletere* = "east." Likewise, if C were in position 3 or 5, C would be facing south, which would make *északra* = "south," which is incorrect. And if C were in position 6, C would have a single stone to the east, so *nyugatra* = "east," which again is forbidden. By a process of elimination, we can conclude that C must be in position 2, facing north. So *északra* = "north," and *nyugatra* = "west."

Now that we have all the translations, we can complete the rest easily:

Let's look at what B is saying:

> "South (and to my left) are no stones"
> "North is one stone"
> "Behind me are two stones"

So B is in position 4, facing west.

D is saying:

> "West (and to my right) are two stones"
> "North is one stone"
> "Left is no stones"

This puts D in position 5, facing south.

		A←	•			
		•	•	•		
•	C↑		•			•
B←		•		•		D↓
			•			

100 LANGUAGE LAVA

Pita lives at A and Sala at B.

The answer a) is correct.

Every sentence is of the form [Name] *pera kana* X *ieno*, [Name] *pera kana* Y *ieno*, where X and Y are one of *auta*, *ilau*, *ata*, and *awa*. It would seem likely that these four words indicate relative directions. (The phrase *pera kana* means "house" and *ieno* means "is located.") You know that Manam islanders do not use compass directions, so *auta*, *ilau*, *ata*, and *awa* are *not* permutations of north, east, south, and west.

This is a very difficult problem, but you may have noticed from the map that:

i) The relative direction between Mombwa and Kulu is the same as between Tola and Sulung.

ii) Onkau, Tola, and Sulung are the three most inland houses, and in the sentences with an *auta* and an *ilau* they are always referred to by *auta*.

The answer is that Manam islanders describe relative directions based on distance from the sea, and rotational direction around the island. The meanings of the words are as follows:

auta: inland/upland
ilau: seaward
ata: clockwise around the island
awa: counterclockwise around the island

Arongo's house is counterclockwise (*awa*) and seaward (*ilau*) from Sulung, so a) is correct.

The Manam system is familiar to mathematicians. It's equivalent to the use of "polar coordinates," which are a way of describing the position of a point in a plane based on its direction from a reference point, or pole, and its angle from a reference direction. The pole in the Manam system is the center of the island, the crater of the volcano. The relative positions upland/seaward describe distance from the crater, and the relative positions clockwise/counterclockwise describe the rotational angle around the crater. The system only works because the island is almost perfectly circular.

Manam is one of only three known places in the world that use a polar coordinate system for relative directions. The other two are Makian, a volcanic island in Indonesia, and the Fijian island of Taveuni.

Appendix

Pronunciations of some of the curious phonetic symbols used in this book.

ABKHAZ, introduced on page 63

ə like the "a" in "about"

ʒ like the "s" in "leisure"

ɕ like "sh" in "ship," but with the flat of the tongue, rather than the tongue tip, against the hard ridge behind the gums

ʁ similar to a French "r," but not rolled/trilled. Like the "ch" in "loch," but with the tongue even further back; the back of the tongue touching the uvula, and with vibration of the vocal folds.

NAVAJO, page 64

ł Put your tongue on the roof of your mouth as if you are about to say an "l," and then blow. In other words, an "l" with no vibration of the vocal cords. This is a common sound in Welsh, where it is spelled "ll."

ʔ the glottal stop—i.e., the sound in the middle of "uh-oh"

ǫ like "o," but nasalized, similar to the "-on" in the French word *chanson*

ą like "a," but nasalized, as in the French *en*, for "in"

CHALCATONGO MIXTEC, page 66

č   like "ch" in "church"

ž   like "s" in "leisure"

ʔ   as above in Navajo

AMHARIC, page 95

ə   as above in Abkhaz

ḥ   like the English "h"

ḫ   like the English "h"

ṗ   "p" with a little popping sound

A List of the Puzzles and Their Sources

The Lingo Bingos *Vocab Test A–L* and *Vocab Test M–Z* were written by Chris Maslanka. They originally appeared in his brilliant *Guardian* puzzle column.

All the other Lingo Bingos were written by me. The map quoted in *Polish Phonetic Spelling* is taken from *The Red Atlas* by John Davies and Alexander J. Kent, University of Chicago Press, 2017. All the data for *Animal Sounds* was taken from the database of animal sounds at the University of Adelaide School of Electrical and Electronic Engineering. The translations of the Universal Declaration of Human Rights in *Special Letters* are all taken from omniglot.com.

Below is a list of all 100 puzzles and their authors. If the puzzle originally appeared in a linguistics olympiad, I mention which one and, where I was able to find out, the year it appeared. If a puzzle appeared in more than one olympiad, I provide the name of the competition where I first spotted it.

The abbreviations are:

UKLO: United Kingdom Linguistics Olympiad
NACLO: North American Computational Linguistics Open
 Competition
OzCLO: Australian Computational and Linguistics Olympiad
AILO: All Ireland Linguistics Olympiad

Panini: Panini Linguistics Olympiad, for Indian schoolchildren,
    named after Panini, the ancient Sanskrit grammarian
Brazil: Olimpíada Brasileira de Linguística
Netherlands: Taalkundeolympiade
IOL: International Linguistics Olympiad

I adapted most of the problems to make them suitable for a general
readership, and usually gave them a new title.

I have attempted to contact all the authors. If you are one of the few
I was not able to reach, please contact the publisher.

INTRODUCTION
    Japanese "counting words"
    puzzle
    Harold Somers, AILO 2014

1. ODD COUPLES
    *Problems from Linguistics
    Olympiads 1965–1975*, editors
    V. I. Belikov, E. V. Muravenko,
    and M. E. Alexeev, MTsNMO,
    Moscow, 2007

2. ICE CHEESE
    Patrick Littell, NACLO 2014

3. IT STARTED TO RAIN
    Sadid Hasan, OzCLO 2018

4. THE BAD TRANSLATION
    Harold Somers, UKLO 2011

5. THE WORLD'S FUNNIEST
    CROSSWORD
    Jordan Ho, Patrick Littell,
    NACLO 2017

6. WHO DO YOU THINK
    YOU ARE?
    Alexander Piperski, IOL 2015

7. WHAT'S MY NUMBER IN
    ITALY?
    Alexander Piperski, NACLO
    2012

8. EMBED WITH A LINGUIST
    Monojit Choudhury, Panini
    2018

9. A CROMULENT
    CONUNDRUM
    Dragomir Radev, Christiane
    Fellbaum, Jonathan May,
    NACLO 2015

10. WE COME IN PEACE
    Kevin Knight, Simon Zwarts,
    UKLO 2012

30. CELTIC COUNTING
Pedro Neves Lopes, Brazil 2013

31. GRIPPING REEDS
Taken from *Reading the Past*,
British Museum Press, 1990

32. SHUT UP, SON!
Martin Worthington, UKLO
2020

33. THE KINGS OF OLD
PERSIA
*Problems from Linguistics
Olympiads 1965–1975*

34. CHAMPAGNE FOR
CHAMPOLLION
Tom Payne, UKLO
Breakthrough Workout

35. DEATH ON THE NILE
Adapted from *Lost Languages* by
Andrew Robinson, Thames &
Hudson, 2009

36. PURPLE REIGN
Harold Somers, UKLO 2013

37. CRETAN CRUNCHER
Author unknown, Estonian
Linguistics Olympiad 2007

38. MASTERS AND SLAVES
Todor Tchervenkov, NACLO
2007

39. AND THE OSCAR FOR
OSCAN GOES TO . . .
Michael Salter, UKLO 2018

40. NORSE CODE
Catherine Sheard, UKLO

41. THE FARFAR NORTH
Yu. Kuznetsova, Panini 2011

42. MY ROMAN FAMILY
*Problems from Linguistics
Olympiads 1965–1975*

43. MEET THE RELATIVES
Babette Newsome, UKLO 2017

44. RICE WITH THE
GRANDKIDS
A. Kretov, Panini 2011

45. THE COUSIN WHO
HUNTS DUCKS
Emily Bender, NACLO 2015

46. AMY, SUE, AND BOB, TOO
John Mansfield, OzCLO 2019

47. MY WIFE'S FATHER'S
MOTHER'S BROTHER
Alan Chang, NACLO 2013

48. BURMESE BABIES
Ivan Derzhanski, Maria Cydzik,
IOL 2009

Acknowledgments

This book would not have been possible without the generous support of the linguistics olympiad community. I first approached Dick Hudson, Emeritus Professor of Linguistics at University College London and chair of the UKLO committee, with the idea for a book of olympiad puzzles in the summer of 2019. He liked the idea right away and has been extremely kind and helpful all the way through the project.

All the national organizations I approached made their problem archives available to me. Thank you, Dragomir Radev from NACLO, Harold Somers from AILO, Mary Laughren from OzCLO, Monojit Choudhury from the Panini Linguistics Olympiad, and Bruno L'Astorina from the Olimpíada Brasileira de Linguística. Harold was especially generous in helping proofread the manuscript.

One of Dick's first recommendations was to introduce me to his crack team of champion linguists: Sam Ahmed (IOL gold medalist 2015, '16, '17), Liam McKnight (IOL gold medalist 2015, '16, '17, '18) and Elysia Warner (IOL gold medalist 2014). These three super-smart hall-of-famers were brilliant guides to the olympiad archives, helping to direct my research, as well as patiently explaining how to solve several problems that had me tearing my hair out.

I'd like to thank all of the authors of the puzzles for letting me use or adapt their material, with a hat-tip to Patrick Littell, Harold Somers (again!), Babette Newsome, Bruno L'Astorina, and Alexander Piperski, who between them devised more than a third of all the problems. Listed below by chapter are the experts, and native speakers, who helped me check the puzzles and answer my questions.

1. **Ok-Voon Ororok Sprok:**
   Sadid Hasan, senior director of Artificial Intelligence at CVS
   Monojit Choudhury, who as well as helping run the Panini
      Linguistics Olympiad is a linguist at Microsoft
   Christiane D. Fellbaum, professor of linguistics at Princeton
      University

2. **Celts, Counts, and Coats**
   Damian McManus, professor of early Irish at Trinity College,
      Dublin
   Gareth Ffowc Roberts, emeritus professor of education at Bangor
      University
   Graeme Trousdale, professor of linguistics and English language
      at Edinburgh University
   Mart van Baalen, Dutch speaker
   Peter O'Donohue, York Herald at the College of Arms
   Maxime Rovère, French philosopher who was once a visiting
      professor at the University of Papua New Guinea in Port
      Moresby

3. **All About That Base**
   Eleanor Robson, professor of ancient Middle Eastern history,
      University College London
   Sabine Hyland, professor of world Christianity, St. Andrews
      University
   Gary Urton, professor of pre-Columbian studies at Harvard
      University
   Andrew Pawley, emeritus professor of linguistics at the Australian
      National University
   Mary Walworth at the Max Planck Institute for the Science of
      Human History

Robert Teare, Manx Education Language Officer at the Manx
Language Unit

4. Decipher Yourself!

Martin Worthington, senior lecturer in Assyriology at Cambridge
University

Jonathan Taylor, assistant curator, Cuneiform Collections, at the
British Museum

Katherine McDonald, senior lecturer in classics and ancient
history at Exeter University

5. Relative Values

Adam Kuper, visiting professor of anthropology at the London
School of Economics

Robert Parkin, emeritus fellow of anthropology at Oxford
University

Ileana Paul, specialist in Malagasy at the University of Western
Ontario

John Mansfield, lecturer in languages and linguistics, Melbourne
University

Carmel O'Shannessy, senior lecturer in linguistics at the
Australian National University

Ásgerður Harriss Þuríðardóttir Jóhannesdóttir Johnsen (her full
name), my Icelandic friend who kindly called up the Icelandic
Naming Committee for me

6. Aiding and Alphabetting

Wendy Osmond, stenographer, wendyosmond.co.uk

Father Joseph at the Mount Saint Bernard Abbey, Leicestershire

Miguel Neiva, the designer who invented ColorAdd

7. Treacherous Subjects

Sammy Andersson, linguistics PhD at Yale University

Patrick Littell, researcher with the Canadian federal government, whose main role is helping support language technology R&D in Indigenous languages

Katie Sardinha, independent researcher into Kwak'wala

Emma Felber, Quechua speaker

Matthew S. Dryer, professor of linguistics at the State University of New York at Buffalo

8. Games of Tongues

Alexey Pegushev, Esperanto speaker

Margareta Jennische, president, Blissymbolics Communication International

Yukio Ota, inventor of LoCoS

Sonja Lang, inventor of Toki Pona

The Toki Pona Facebook Group

9. Script Tease

Candessa Tehee, coordinator, Cherokee Language and Cherokee Cultural Studies Programs, Northeastern State University

Kevin Russell, professor of linguistics at the University of Manitoba

Patrick Littell (see 7 above)

Coleman Donaldson at the University of Hamburg

Sirtjo Koolhof, independent scholar and Bugis speaker

Chris Percy, Mandarin speaker

10. Oh My Days

Girma Tadele, Amharic speaker

Harri Kettunen, lecturer in the Department of Cultures at Helsinki University

Stephen Houston, professor of anthropology at Brown University

Brent Woodfill, visiting professor at Georgia State University

Nykkie Woodward, Maltese speaker

Kelmet il-Malti, Maltese Language Facebook Group

Daniel Harbour, professor of the cognitive science of language at Queen Mary University of London; and Bernard Comrie, professor of linguistics at the University of California Santa Barbara, were very kind to provide me with background information about language and writing systems. I'd also like to thank Svetlana Burlak, Sam Cartmell, Dora Demszky, Angikar Ghosal, Tim Grant, Claire Hardaker, Erica Hesketh, Tamila Krashtan, Itziar Laka, Juliet Law, Kate Lupson, Ben Lyttleton, Tom McCoy, Annette Mackenzie, Danylo Mysak, Vica Papp, Chris Percy, Marie Phillips, Michael Salter, Tyler Schnoebelen, and my parents, Ilona Morison and David Bellos.

At Guardian Faber, my fantastic editor Fred Baty heroically wrestled with solving the puzzles as well as the logistical puzzle of publishing such a complicated book during lockdown. Laura Hassan gave overall editorial advice and Marigold Atkey gave valuable early feedback. The fabulous cover was designed by Jack Dunnington, and the excellent illustrations by Andri Johannsson. Thanks to my US editor, Karen Giangreco, and to The Experiment team.

The challenges in editing, typesetting, and checking a book that includes dozens of languages, exotic scripts, and illustrations were formidable. Thankfully, the UK edition counted on an outstanding team: eagle-eyed Ben Sumner, who gave a masterful copy edit; Richard Carr, whose design makes the book look so elegant and enticing; and Ian Fitzgerald, who calmly and professionally held the whole thing together.

I'm lucky to be represented by Rebecca Carter at Janklow & Nesbit, and her colleagues Emma Leong, Kirsty Gordon, Ellis Hazelgrove, and Emma Parry.

Most of this book was written during the coronavirus lockdown. Without the kindness and understanding of my wife, Natalie, I would not have been able to write it at all.

Finally, I'd like to mention my nephew Joshua, who spent two years living in South Korea as an English teacher. 감사합니다 for checking puzzle 90.

Index of Languages

This list includes writing systems and number systems.

About the Author

ALEX BELLOS holds a degree in mathematics and philosophy from Oxford University. His bestselling books *Here's Looking at Euclid* and *The Grapes of Math* have been translated into more than twenty languages and were both shortlisted for the Royal Society Science Book Prize. His puzzle books include *Can You Solve My Problems?*, *Puzzle Ninja*, *Perilous Problems for Puzzle Lovers*, and *Puzzle Me Twice*, and he is also the coauthor of the mathematical coloring books *Patterns of the Universe* and *Visions of the Universe*. He has launched an elliptical pool table, LOOP. He writes a popular-math blog and a puzzle blog for *The Guardian*, and he won the Association of British Science Writers award for best science blog in 2016. He lives in London.

alexbellos.com | 🐦 alexbellos

Enjoy hours more fun (and some torment!) puzzling with Alex Bellos . . .

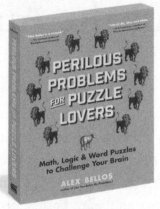

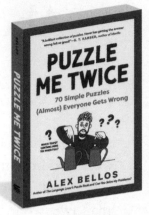

Paperback | 352 pages
$16.95 US | $21.95 Can.
978-1-61519-388-2

Paperback | 288 pages
$16.95 US | $21.95 Can.
978-1-61519-718-7

Paperback | 224 pages
$16.95 US | $21.95 Can.
979-8-89303-028-0

Then awaken your inner math artist in his coloring books with Edmund Harriss!

Paperback | 144 pages
$14.95 US | $19.95 Can.
978-1-61519-323-3

Paperback | 144 pages
$14.95 US | $19.95 Can.
978-1-61519-367-7

Available wherever books are sold